VizAbility™

Handbook

PWS Publishing Company
20 Park Plaza, Boston, Massachusetts 02116 – 4324

International Thomson Publishing
The trademark ITP is used under license.

The PWS VizAbility team: Jonathan Plant,
Patrick Boles, Ellen Glisker, JP Lenney,
Ken Morton, Ed Murphy, and Helen Walden

Design/Production: MetaDesign San Francisco
Cover Design: Terry Irwin, Jeff Zwerner

Printed and bound in the United States of America
95 96 97 98 99 – 10 9 8 7 6 5 4 3 2 1
ISBN 0-534-95112-0

Copyright © 1996 by PWS Publishing Company,
a division of International Thomson Publishing Inc.

All rights reserved. No part of this book may be reproduced,
stored in a retrieval system, or transmitted, in any form or by
any means – electronic, mechanical, photocopying, recording,
or otherwise – without the prior written permission of PWS
Publishing Company.

For more information, contact:

PWS Publishing Company
20 Park Plaza, Boston, MA 02116-4324

Nelson Canada
1120 Birchmount Road, Scarborough, Ontario,
Canada M1K 5G4

International Thomson Publishing Europe
Berkshire House 168-173, High Holborn,
London WC1V 7AA, England

International Thomson Editores
Campos Eliseos 385, Piso 7, Col. Polanco,
11560 Mexico D.F., Mexico

Thomas Nelson Australia
102 Dodds Street, South Melbourne,
3205 Victoria, Australia

International Thomson Publishing GmbH
Konigswinterer Strasse 418, 53227 Bonn, Germany

International Thomson Publishing Asia
221 Henderson Road, #05-10 Henderson Building,
Singapore 0315

International Thomson Publishing Japan
2-2-1 Hirakawacho, Chiyoda-ku, Tokyo 102, Japan

Adobe, Adobe Illustrator, Adobe Photoshop, and PostScript
are trademarks of Adobe Systems Incorporated. Apple
and Macintosh are trademarks of Apple Computer, Inc. Canon
is a trademark of Canon, USA. Clarendon and Swift are
trademarks of Linotype-Hell, AG. Debabilizer is a trademark
of Equilibrium Technologies, Inc. ITC Officina Sans is a
trademark of International Typeface Corporation, Inc. LaserJet
is a trademark of the Hewlett-Packard Company. Windows
and Microsoft Word are trademarks of Microsoft Corporation.
VizAbility is a trademark of PWS Publishing. Blockbuilder © 1995
The Board of Trustees of the Leland Stanford Junior University.
QuarkXPress is a trademark of Quark, Inc.

VizAbility™

Handbook

Kristina Hooper Woolsey

PWS Publishing Company

An International Thomson Publishing Company

Boston • Albany • Bonn • Cincinnati • Detroit • London • Madrid
Melbourne • Mexico City • New York • Paris • San Francisco
Singapore • Tokyo • Toronto • Washington

Table of Contents

Preface

Imagine a cube... Each side represents one element of your natural visual abilities.

Together they can change the way you think, learn, and create.

Learn to see and draw what you see... Learn to imagine and diagram your thoughts...

Consider environments that support visualization, and enter into a culture that encourages invention...

Develop your VizAbility...

And learn to take your ideas from the realm of possibilities into the world of reality. **011 Introduction**

Preface

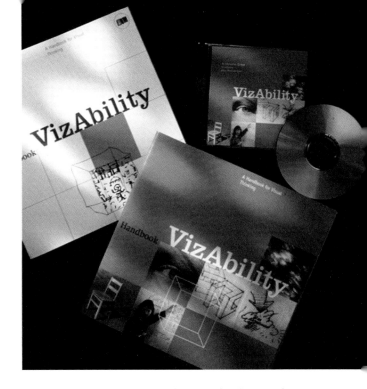

Your natural visual abilities can bring greater productivity and inventiveness to every aspect of your life. For if you can make your ideas visible, you can harness their power.

You can see their strengths and weaknesses more clearly, and explain them to friends more easily. You can complement your other mental skills – verbal, analytic, artistic, interpersonal – with visualization techniques to extend how you think, learn, and create.

The primary goals of *VizAbility* are:

To familiarize you with the visual culture
To make you aware of your own visual abilities
To exercise and improve your skills in visualization
To incorporate these skills into your daily life and professional activities.

These goals are accomplished in two fundamental ways: exposure and training. You will have the opportunity to observe people who are part of this culture, to listen to their ideas and explore their workspaces. You will also be provided with training in specific areas – seeing, drawing, diagramming, and imagining – in order to hone the skills that are critical to participation in this culture.

Although all three *VizAbility* components can be used on their own and still provide valuable insight and experiences, they are designed to interact with each other. Their intent is to offer the medium – book, computer, or paper – that best accomplishes an understanding of the material. *VizAbility* is designed to encourage cycles of reflection and action, abstract ideas and concrete examples, through alternating experiences of reading, hands-on computer activities, and off-screen sketching on paper.

The *Handbook* is designed to provide you with a broad conceptual overview of visual thinking and to guide you through the interactive experiences of the *CD-ROM*. Its structure parallels that of the *CD-ROM*, and provides commentary on the various exercises. You will be able to "jump" directly to sections on the *CD-ROM* that are referred to in the *Handbook*, by using your computer keyboard. (For Macintosh users, use Command ###; for Windows users, use Alt ###.) These sections are indicated throughout the *Handbook* by three numbers.

For example, Command **331** accesses the introduction to HIDDEN PICTURES in the SEEING section of the *CD-ROM*. The visual index on page 192 lists all of these jump points.

The *CD-ROM* offers an interactive environment in which to challenge and improve your visual skills. It includes video-based interviews with visual thinkers, interactive exercises to try by yourself or with others, and galleries of examples created by other visual thinking students. These direct experiences form the basis for the reflections found in this *Handbook*.

The *Sketchbook* offers you a forum for recording your ideas and experiences, including the paper-based exercises found both in the *Handbook* and on the *CD-ROM*. As you progress through the product, and as you experiment and reattempt visualization activities, your *Sketchbook* will become a map of your learning. Many visual thinkers refer to such books as "idea logs" and carry one at all times. We encourage you to do the same as you immerse yourself in the processes of visualization.

The Context for This Product

VizAbility is an outgrowth of a number of traditions. Central to its approach is Robert McKim's book, *Experiences in Visual Thinking*, and the visual thinking course that he developed and taught at Stanford University's Mechanical Engineering Department.

There are a number of other San Francisco Bay Area individuals and groups that, in the last twenty years or so, have also influenced the concepts presented in this product. These include the Mathematics Imagery Group at UC Santa Cruz, which investigated the relationships between mathematics, imagery and computers in 1976; David Sibbet and Fred Lakin, both of whom have been pioneers in the use of hand-drawn and computer-generated graphics for idea facilitation; the Computer Science Department at Stanford University and Xerox PARC, which have provided stimulating environments in which to explore ideas about visualization and its applicability in computer worlds; and the Apple Multimedia Lab, which provided a supportive environment for the development of interactive ideas and prototypes between 1987–1992.

I am indebted to the contributions these people and environments have made, and hope that this product will effectively extend and reflect the quality of these earlier traditions.

Why VizAbility?

"VizAbility" is the term coined in this product to describe a set of human capabilities that are pervasively visual and spatial.

Other terms that have been used to describe similar approaches include visual thinking, non-linear thinking, lateral thinking, spatial reasoning, right brain thinking, and design competency.

The more general and innocuous "visual ability" phrase has been chosen here for a number of reasons. For one, it avoids some non-productive "either-or" arguments such as left brain vs. right brain, thinking vs. feeling, art vs. design vs. science, and linear vs. non-linear. It has been selected because it is inclusive instead of exclusive, and explicitly acknowledges the multiple modes of representation used in cognitive activities, while at the same time emphasizing a set of representation that is typically "invisible" in formal academic training.

It has also been carefully chosen to avoid some prominent, sticky, theoretical issues of neurobiology or psychology, and to take instead a pragmatic approach toward developing a set of effective skills.

The Production Team

VizAbility has drawn on the talents of a very large number of people. Robert McKim encouraged its early development and provided important insights above its form. Scott Kim and Gayle Curtis helped me bring the concepts together and to complete a pre-production phase on the overall project. MetaDesign supervised all production activities and created the product's visual design. Bill Purdy acted as producer of the *CD-ROM*, and Cindy Rink served as co-producer. Other key individuals in the production process have included Wendy Slick, Bill Hill, Terry Irwin, Jeff Zwerner, Don Ahrens, Marabeth Harding, Jim McKee, Helen Walden, Jym Warhol, and Hawkwan Warhol.

An equally large number of people have enhanced the contents of this product by contributing their interviews. This "cast of characters" includes Denny Boyle, Alison Quoyeser, Rob Semper, Frank Wiley, Primo Angeli, Terry Irwin, David Sibbet, and Scott Kim. I thank all of these individuals for their insights and their generosity in sharing them.

I have been given the task of writing this *Handbook*. It has benefitted greatly from the overall product development and the insights of my many colleagues on this project. Cindy Rink has been particularly helpful as my editor on both the *CD-ROM* and this *Handbook*.

PWS Publishing Company has provided unending support and enthusiasm for this project. They have acknowledged all the risks of this new kind of medium, yet have never swerved in their collaboration and cooperation. Special credit for this perspective goes to Jonathan Plant, J.P. Lenney, and Ed Murphy.

I thank everyone involved for helping to bring this product from a vague but important idea to a concrete product.

Kristina Hooper Woolsey 1995

Introduction

"You have a lot of ideas stuck in your head. Let them out."

People in a variety of fields recognize that our increasingly complex world requires new forms of communication and problem solving. They have discovered that visual representations can be powerful aids in accomplishing new modes of innovative thought.

The visual thinking culture introduces a methodology as well as an awareness. It is about problem finding as well as problem solving. It is about maintaining momentum on a task by finding better ways to do things, based on direct concrete experiences. It is about the belief that everyone is a designer. And it is about knowing how to sustain "flow experiences," that state of productivity between boredom and anxiety.

Your mind doesn't stop, ever. It is important to have fluid representational skills that can stimulate, keep pace with, and express your image-rich thoughts. To address this, tools and methodologies are being developed that allow doctors, scientists, students, and business people – as well as designers, artists, and engineers – to express their ideas visually. New computer-based media are simplifying the integration of imagery into everyday dialogues and presentations. The casual electronic tools that bring these skills to the forefront of our culture are already on their way.

VizAbility is designed to help you to be ready for these developments, to participate in their use.

At some point, every person must take charge of their own thinking, learning and education. This product can help you take advantage of the many resources and perspectives on visual thinking that are available to you. It offers you a peek at the possibilities, so that you can enhance your own natural visual skills and take advantage of them.

Much thinking occurs below the level of conscious awareness. People come up with a solution suddenly, often after they are stuck on a problem for days. If you can control your imagery, you can take advantage of the thinking that goes on below your level of consciousness and bring good ideas into concrete form.

Diverse Audiences

Who can benefit from *VizAbility*? Everyone who needs to imagine, define, resolve, invent, analyze, and communicate.

The audience ranges from the mathematician who wants to imagine flow phenomena, to the geneticist who wants to consider the structure of a molecule, to the business executive who needs a clearer mental map of his company's organization, to the writer who wants to visualize the sounds and the colors of a scene before writing about it.

Visualization techniques can be absolutely central to art and design professionals, individuals who typically work in highly visual environments.

They can also be used by scientists and other fundamentally analytic professionals – such as physicists, engineers, and statisticians – to formulate a problem and represent its solution to colleagues.

The audience for this product encompasses people who are learning to communicate their ideas – the graduate student writing a Ph.D. thesis, the politician trying to position himself, the student taking an exam, or the architect working on the design of a new house. Teachers can broaden their skills set by using visual thinking techniques for themselves and their students. Even fundamentally verbal professionals – such as speech writers, broadcasters, and newspaper journalists – can benefit from visualization. It can help them to formulate their words, or to combine them with sketches or other imagery, in order to lend greater coherency and depth to their ideas.

In short, visualization can be useful to everyone, and to every kind of activity, from solving problems in a professional arena to cooking for one's family, from managing personal relationships to new products. It runs the gamut from the practical to the esoteric, the abstract to the concrete.

The VizAbility Cube

The model for the *VizAbility* product is a cube. This six-sided shape provides a convenient framework in which to represent different elements of the visual culture and your visual abilities. Like a cube, this product has a non-linear form that can be approached and enjoyed from any direction—a metaphor that coincides with the exploration-based nature of multimedia. However, there are certain foundational elements embedded in *VizAbility* that invite directional flow. They form the basis for this Handbook's six sections, as follows:

Environment

The foundation of visualization is a supportive environment. Ideally, this environment consists of a benign physical setting formed of light, air, and space; an array of useful tools and materials, and the companionship of non-judgmental colleagues who provide positive, generous expectations and whose attitudes nurture inventiveness.

The Environment facet of the *VizAbility* cube is designed to help you:

• Experience the environments of some visual thinkers.
• Become self-reflective about your own environment.
• Learn to establish a productive working environment.
• Understand how objects in your environment can stimulate ideas.

It is also your opportunity to meet *VizAbility*'s primary cast of characters, people who approach their environments and their lives from a visual perspective.

Culture

The culture that inspired *VizAbility* is one that uses visual representations to invent new ideas or to stand in for developing ideas, much as a variable in algebra stands in for a relationship until the details about a particular situation are worked out.

It is a culture of contrasts—abstract and concrete, personal and public, informal and formal. It encourages a set of tools and techniques that aid the visualization process. These include: sketching as a method of communication; "idea logs" as places to capture and record ideas; prototyping as a way to experience and share ideas in three-dimensional form; and critiquing as a process that develops and refines concepts.

In addition, it stresses the importance of mental and physical readiness in order to achieve a state of optimal creative "flow."

The Culture facet of the *VizAbility* cube is designed to help you:

• Gain further insight into the visualization culture.
• Learn about some of its tools and techniques.
• Understand the essential cycle of sketching, prototyping and critiquing.
• Experience some activities that ready you for visual thinking.

Seeing

The third facet of *VizAbility* focuses on tuning up your seeing skills. Visual exercises stretch our ability to see, they increase our mental and visual flexibility, and provide multiple perspectives on even common everyday objects.

The Seeing facet of the *VizAbility* cube is designed to help you:

• Become aware of how you see.
• Unblock your visual stereotypes.
• Translate motion into form.
• Learn to notice details.
• Sort, categorize, and group elements visually.
• Experience how differently people view the same thing.

Drawing

In the context of visual thinking and communicating, drawing does not require what is commonly described as "artistic ability." Instead, it is used to help you become familiar and fluent with a basic tool of expression. Drawing is a skill that enlivens the imagination, that brings ideas into the material world for conversation and debate, and that creates simple representations of the objects and scenes we have experienced.

The Drawing facet of the *VizAbility* cube is designed to help you:

• Use drawing to enhance seeing.
• Learn a variety of basic drawing techniques.
• Gain an intuitive understanding of shading and perspective.
• Learn how to construct simple objects.
• Experience diversity in drawing approaches and competencies.

Diagramming

Diagrams provide a format for making even abstract ideas concrete enough to share with others. They allow you to hold pictorial conversations, to record events over time and space, and to sketch out interrelationships. Diagrams let you describe and illustrate an idea whose details are still in progress, thereby inviting the input and collaboration of others.

The Diagramming facet of the *VizAbility* cube is designed to help you:

• Familiarize yourself with diagrammatic forms.
• Learn how to use and create symbols.
• Gain skill in rendering ideas.
• Appreciate diagramming in everyday contexts.

Imagining

Our inner source of visualization is our imagination. Generating ideas and working with vivid mental imagery are the core experiences of the sixth face of the cube. A rich repertoire of spatial puzzles, audio environments, and brainstorming techniques is offered to help you build fluidity and facility in the production of new ideas.

The Imagining facet of the *VizAbility* cube is designed to help you:

- Enhance inner visualization.
- Recognize the power of your own imaginative abilities.
- Use collective imagining to solve problems.
- Articulate ideas through sound and imagery.

A Cubic Overview

The facets of the *VizAbility* cube, like the visual culture itself, form different dichotomies. Environment, Drawing, and Diagramming reflect the external, public, concrete world. Culture, Seeing, and Imagining address the inner, personal, cognitive domain.

Environment is about physical place, Culture is about collaborative attitudes. Seeing and Drawing are perceptual, Imagining and Diagramming are conceptual. The interactions of these can be powerful.

VizAbility is mind and media in dialogue. It is about the intersection of actions and thoughts, the transitions between our minds and the world around us. It is about the verbs between the nouns. Your visual abilities are the conduit that carries your ideas from one to the other.

VizAbility is not just about making images or looking at them, about how one can use an image to understand or explain something. It is about how a series of images can show a changing idea and evolve into practical, unambiguous products.

Today most of this is rather opaque.

Environments are things most people take for granted, which come to attention only when they are especially problematic. Cultures are invisible, especially to their members, until one becomes aware of them. Seeing seems mundane to most. Drawing is a thing people are afraid of, a domain reserved only for the professionals and the artists. Diagrams are considered useless, and their surface forms are confusing to many. Imagination is thought to be frivolous.

Hopefully as you and others experience this product, these points of view may change.

Environments can become central supports for
innovative work, full of inspiration and tools and
materials. The visual culture can become a prominent
contributor to effective and realistic ways to work.
Seeing can be a highly developed and enjoyable
skill for many. Everyone can draw and bring their
ideas about the world to the table for analysis.
Diagrammatic fluency can help people give form
to their abstract ideas. Imagination will be widespread, cultivated, treasured, and benefitted from.

Visualization can complement other modes of
thought and representation, and each of us will be
better prepared to address that which we engage.

This *Handbook* will address these and other issues.
The cube will be rolled and each of the six faces
will offer up a set of observations and experiences.
We hope they will provide you with new ways of
looking at both your inner and outer worlds.

Environment

It is the physical space around you that can encourage your own visual abilities.

You can surround yourself with air and light and large surfaces.

You can organize your visual tools—sketchpads, whiteboards, posting spaces, computer applications—
for flexible and spontaneous use.

...fill your space with engaging objects, which can stimulate ideas and provoke
new and old associations.

Visit four people who have created spaces to support and enhance visual thinking.

Learn new perspectives from them in order to prepare your own environment. **101 Introduction**

Introduction

The more aware you become of your visual abilities, the more important your environment will become.

You will find that the quality of light matters as you learn to focus your seeing. You will find that you need space, on your desk or at a table, in order to sketch or diagram ideas on large pieces of paper. You will learn how the sounds surrounding you can help to focus your attention and revitalize your energy. You will begin to notice the objects around you and to use them as resources for thinking, sketching, and communicating ideas to others.

You will probably start to build a supportive environment of colleagues and friends who can think visually with you – people who are responsive as you sketch out your ideas on a napkin over lunch or at a whiteboard in the office. Colleagues who respect the content of your thoughts, and don't laugh at or demean the drawings you use to explain your ideas. People who are ready to collaborate, to critique, to define and solve problems in visual ways.

Of course, it is possible to see, imagine, diagram, and draw without a supportive physical and social environment. But if you can establish this foundational element in your life, you can take energy from it. You can create a healthy visual culture around you that is positively "contagious," instilling a new excitement in both your work habits and your productivity.

Such visual environments do already exist. The design community in particular – architects, product designers, and graphic artists – creates vibrant and highly sensory environments in which people can work. Painters and sculptors surround themselves with light, space and stimulating objects to encourage their own creativity. Engineers and computer scientists employ whiteboards and collaborative brainstorming as tools for ideation.

66 The environment to me is probably the single most important element in creating visual work. It's the place where thinking occurs. And the control of that environment...is to me the most important aspect of being creative and taking advantage of my creativity." - Bill Hill 101 Overviews

It is no accident that, in many of the places where innovation is taking place, the visual culture pervades! Unfortunately, however, these creative environments are typically isolated domains.

The grand opportunity is to extend this visual culture, so that it encompasses more general communities, and encourages and supports fresh ideas everywhere. The visual culture should, for example, become central to our schools. It should pervade our mathematics classes and our creative writing classes, not only in kindergarten and primary school, but at all levels of education. Scientific laboratories, publishing offices, and other places requiring thoughtful work could thrive with this type of supportive environment. Both our homes and our offices could benefit from this fresh perspective.

Creating Your Environment

People create their own physical and social environments. Sometimes this is accomplished by default, such as when individuals consider themselves powerless to structure a supportive environment, or when they give this task to others to achieve, or when they consider it irrelevant to their daily activities. Yet if you are to make full use of your visual abilities, you need to consider your environment as a critical foundational element. For when you choose to engage your visual abilities in perceptual and conceptual tasks, you also choose to deal with the material world – how it looks, how it supplies you with tools, and how it inspires your ideas.

To make this more concrete for you, we have visited four individuals who use visualization frequently in their daily activities and who work in different environments – an office, a classroom, a public space and a studio. As you will come to notice, there are great similarities between these individuals due to their visual orientation; they use different environments very self-consciously to visually support themselves and their colleagues, be these clients, students, visitors or co-workers.

These individuals are not from the most obvious visual fields, either. They are not working artists, for example, but instead represent the domains of engineering, teaching, science, and software design.

By visiting them you can directly experience how their ideas and the objects with which they surround themselves interact. You can begin to experience the fundamental tension of visual cultures – between the real and imagined, the concrete and the abstract, the things and the ideas. And you can begin to acknowledge and appreciate the opportunities created by this tension.

You can meet these individuals in the Environment section of the *CD-ROM*. There you can view their workplaces in detail, view the objects and tools they use, and listen to short interviews about their ideas. As you listen to them, notice how their thinking and productivity relates to their environments. Consider how similar or different your workspace is, and how you might borrow some good ideas.

66 I think an environment that allows kids to paint and draw and to build and to make puppets and...to get up in front of the class and act with costumes that they've created – this is the class that is valuing who these kids really are."
-Alison Quoyeser 101 Overviews

Offices: Personal Workspaces

Denny Boyle is an engineer and product designer for IDEO, a product design firm near Stanford University.

He uses a range of visual representations to stimulate his own inventiveness, expressing his ideas on whiteboards, in idea logs, and in three-dimensional prototypes.

Objects in the Office Environment

One of Denny's principle workspaces is his office, but it is not just an everyday "desk/computer/bookshelf" office. True, Denny has a spacious table on which to work and lay things out – but the rest of his environment is a storehouse of inspirational objects, artifacts from earlier projects, and informative or meaningful brochures, articles, and notes.

Denny uses the objects around him to motivate new ideas and reflect upon old ones.

"Sometimes problems just need a push, they need a starting idea...a piece of wire, or a spring, or a material...and if you can grab that quickly, if you don't even have to walk around for it, it sometimes gets you off the dime." **112**

Rather than discard these concrete objects, Denny deliberately collects them in order to recall past histories or to jog his own thinking in a different direction.

He also keeps records of his projects in his drawers and bookshelves. These are principally in the form of "idea logs," sketchbooks filled with drawings and descriptions of evolving products. Denny takes great pride in always having an idea log at hand to capture good ideas; he carries a small 3" x 4" notebook in his pocket at all times, and also maintains larger 8 1/2" x 11" notebooks in which to sketch ideas and attach relevant materials, such as blueprints and memos.

66 I've found that surrounding myself with as much material as I can, whether it's related or not, it all tends to be related somehow, indirectly. I can make some lateral connection between things that aren't normally associated, to make them try and express my idea." - Denny Boyle

In these ways – through inspirational objects and personal sketchbooks – Denny is able to use both his past experiences and the world around him to address new project requirements. He does not rely solely on his own divine inspirations, though surely he welcomes such happy occurrences. Instead he consciously arranges his environment to support inventiveness.

Public and Collaborative Spaces

In addition to Denny's office, which is his personal space, the IDEO facility offers him other elements and environments that are beneficial to visual thinking. He has access to a number of conference rooms lined with whiteboards to support collaborative brainstorming and project presentations, and he also enjoys the company of colleagues who are themselves avid visual thinkers in their own rights.

He also has access to a prototyping laboratory, where he can quickly mock up basic ideas, and collaborate with colleagues in developing more advanced and complex concepts. Given the product orientation of IDEO, the prototyping facility is primarily stocked with mechanical and electrical components.

"You're surrounded by old projects, parts and pieces in bins and boxes...electrical parts, mechanical parts, art supplies of every sort, materials, boards, different pens, cutting instruments. Then there are various shops so that you can cut them, put them together, clamp, glue, paint and lubricate them, and adjust their attitude – with a hammer – if they don't work." Denny Boyle

IDEO's social environment also includes individuals with a range of supporting skills, such as computer-aided design, marketing, and sales. This allows the creation of teams that can liaison with lead designers and address the needs of particular clients.

The importance of collaboration for inventive thinking cannot be overemphasized. When many of us were in school, we thought that we would develop single-handedly all the talents to accomplish great feats. Yet in truth, it is usually the interaction among people with different talents and skills that allows both organizations and individuals to attain their goals. It is important to acknowledge this explicitly when creating a supportive environment for your own visual thinking activities. The visual thinking culture is fundamentally a collaborative one, as most individuals need, and flourish with, social input and outside perspectives.

66 I've referred to idea logs as wallets for your idea currency, a place to record, to sketch things, to keep things in more or less chronological order as they develop on a project... It's surprising how many thoughts you have; they're constant... if you don't get them down you sometimes lose them."
- Denny Boyle

The conference rooms at IDEO follow these principles. They are designed deliberately to facilitate group processes, from in-house brainstorming activities to more formal client presentations. With their ceiling-to-floor whiteboards, they offer an arena where groups can assemble to talk, draw, generate, and critique ideas.

Most effective workplaces have large public rooms available, with group seating, and display and posting areas that can accommodate the development of dynamic conversations and concrete examples. The trick in designing such rooms is to keep them flexible, and to acknowledge that sketches and prototypes of new ideas come in a variety of sizes and shapes.

Today such rooms frequently include computers with projection systems, since computers can store a wealth of graphics almost invisibly. The single screen's best use currently lies in displaying a sequence of examples or illustrating a linear speech.

How does all this relate to you, and your design for a visual workspace? Here are some ideas that can be derived from Denny's workspaces:

- Consider how to surround yourself with concrete representations of your own work so that they can serve as resources to you, both for inspiration and for assistance in explaining ideas to others.

- Be divergent in choosing these objects. A silly toy, a memento of a trip, a snapshot of a friend, or a napkin on which you sketched out an idea, may be just the thing to jog your thinking when you need it most.

- Provide access to prototyping and display areas. Whatever your craft, consider what facilities you need to let you mock up and try out ideas.

- Value your social environment. Regardless of whether colleagues are next door or at the other end of a telephone or electronic mail line, the social fabric you create can be a major contributor to your success. The stimulation of colleagues who have different perspectives and talents can provide the spark for highly productive innovation.

- Try to establish access to public presentation facilities so that you can gather small groups together to consider and critique evolving ideas. Try to equip this room with flexible "show and tell" spaces.

66 We have a shop close by with ten dedicated people that build your ideas once you get them to the stage where it's not productive for a designer to build everything that they come up with." -Denny Boyle

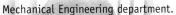

Biography

Denny Boyle is one of the founding design engineers at
IDEO, a consulting engineering and industrial design firm.
He received his BS degree in Mechanical Engineering
and Industrial Design from the University of Notre Dame,
and his MS in Product Design from Stanford University.
Since 1978, he has taught design courses at Stanford's
Mechanical Engineering department.

66 When you try and communicate your idea to someone who
isn't a designer and you sketch it out, usually they'll take
it more seriously if they can see it." - Denny Boyle

Classrooms: Preparing Places for Others to Work

Formerly an architect, Alison Quoyeser is now a second grade teacher. Her diverse background gives her a facility for and experience with complex visual representations.

As a result, she brings to her classroom a special blend of visual skills that engenders learning in small children. She quite naturally encourages the use of objects and pictures to help children bring their own ideas to life.

Although it might seem that the objectives involved in constructing Denny's environment were specifically relevant to product design, we found many similarities between Alison's space and his. Although surface appearances and functions are very different, the same general principles lie behind the design of both spaces.

Visually Oriented Learning Environments
The first impression one gets upon walking into Alison's classroom is that of comfort. It is a well-organized environment for learning. There are bright and lively displays on the walls, many of them made by students. The desks are arranged in small groups to encourage student interaction. There are nooks and crannies that invite young readers and thinkers to linger a while over their work, either alone, in pairs, or in small groups. Materials abound – books, paper, crayons, pens, blocks – and there is a large flat table on which to spread out these materials while working on projects.

66 I think an environment that allows for all kinds of learning styles is really important. It's essential that I come at it from all kinds of ways, since I have all kinds of learners in my classroom." -Alison Quoyeser

Many of us experienced a second grade classroom similar to this. Yet for some reason, we took it for granted that we would all outgrow such environments. If, however, one emphasizes a visual orientation to the learning process, elementary school classrooms are often the better models for learning and inventing – especially in terms of providing materials and tools for project-based work.

Project-based education is currently finding favor in schools, as educators return to the ideas set forth by John Dewey more than fifty years ago. Dewey suggested that active, hands-on experiences are the key to substantive learning. American elementary schools are now applying this kind of constructivist approach to learning and incorporating a range of short- and long-term projects into their curricula. Students are encouraged to become involved, to show what they know, and to explore what they think. Because many of the contexts for these activities are inherently visual and spatial, reading, writing and arithmetic have a new partner in learning – students' visual abilities. These abilities can be developed as a foundational element for traditional disciplines (for instance, the use of blocks and other manipulatives in teaching mathematics) or to encourage new skills (for instance, observational drawing techniques).

This is particularly the case when students use computers in their work. As they begin to incorporate still and moving images in computer-generated projects, they develop skills in basic design and composition, as well as other fundamental visual abilities.

Yet basic training in visual literacy is still not a main curricular element in most school settings. Somehow the American culture continues to believe that visual abilities are "God-given," and that only "talented" people are good at using them. These abilities are more associated with the arts and the vocational crafts than with high level, abstract thinking. The result of this kind of reasoning is a lack of serious training in these skills; they are not considered "trainable," and are not judged to be critical to the basic task of schooling.

Fortunately, however, there are practitioners who have already taken the step towards visual literacy in their classrooms, from elementary school to graduate levels. They have provided students with environments that support analytic and abstract thinking as well as more explicitly artistic tasks.

---

66 In my teaching, if I have something right there that corresponds to what we are talking about, it's such a great aid. And as the years go by, I get better at knowing just what I need around me to reinforce whatever it is we're discussing." - Alison Quoyeser 122

Click around Alison's space to remind yourself what an elementary school classroom looks like, then try to remember what this kind of space looked like when you were a child attending school. Apply what you recall to your current environment.

Objects in the Classroom Environment

Denny uses the objects in his office to inspire himself. Alison fills her environment with objects that can inspire her students. By having easy access to these objects, she can use them in a casual but effective way to stimulate further thought and discussion between herself and her class.

One example is pattern blocks. Playing with these beautifully colored shapes can be a delight in itself for students. With a bit of urging, however, and in the context of lessons, children can use these blocks as valuable aids for learning about symmetry and other basic principles of geometry. Pattern blocks can help in the understanding of fractions, or provide basic counting experiences for beginners. Since there is a generous supply of them, they can also be used to cultivate basic estimation skills, as well as simple multiplication and division abilities.

Alison's classroom contains a range of materials for both "public" and "personal" use. Shelves and closets harbor a wide variety of books, from dictionaries to picture books. Many of the more visible books are thematically related – on topics such as Japan, earthquakes, city planning – and give students further information on topics currently under study. There are also video and audio cassettes, and a number of "cozy corners" around the room that invite students to listen or read by themselves.

Complementing these public materials are construction supplies for personal projects: paper in different sizes and colors, string, paste, beads, glitter, crayons, paints, scissors and staplers – just about anything a student would need to complete a project (including a computer for electronically-generated projects).

The vertical surfaces in the classroom are organized primarily to display these projects. It is not professional materials that surround you in Alison's classroom, but the work of her students. For outsiders, the pattern of these projects shows what is being learned in the classroom. For the students, the displays are visual reminders of their personal accomplishments. They can take pride in their own work and learn new things from the works of others.

The classroom blackboard is a direct parallel to Denny's conference room whiteboards. Here Alison and her students can show their ideas to one another. They can collaborate, discuss, invent, and refine, whether the focus is on a math problem, a sentence structure, or a drawing. The blackboard also displays the scheduled events for the day, to which students can refer regularly.

Many of us remember similar artifacts from our own early education. It is important to purposely recall them as you create your own visual thinking environment, for such classrooms are among the few places in American culture designed to support this kind of learning. The hidden assumption seems to be that

Take a few minutes to compare Denny's and Alison's environments. In your *Sketchbook,* list at least three principles that the two spaces have in common. Include some illustrations/drawings in your list. 113, 123

small children need pleasant visual environments, but that as we get older they become irrelevant. Yet visual thinking is relevant to all ages, and the kinds of environments we see in our elementary schools should not be "grown out of," but, "grown into" for productive and inventive work!

Your Visual Workspace

There are a number of things to learn from this classroom environment that may be useful to you in creating either your own space or a space in which others can learn:

- Concrete materials can teach abstract ideas to children and adults. They can be left out for general "fiddling around" or used in the context of specific lessons; a combination of the two kinds of activities will enhance the effectiveness of the materials. (The Montessori teaching method encourages such use of materials, and in fact provides a wealth of aesthetically and sensually pleasing artifacts – e.g., beautiful woods and bright colors – tailored to present concepts to varied age groups.)

- Spaces need to be provided for project-based work. Reference materials (books, videotapes, pictures) and construction materials (colored paper, poster boards, sticks, string) should be accessible so that students can browse through related materials and then incorporate what they learn into their projects. Basic tools for drawing (crayons, pens, paints) and assembly (glue, staplers) allow students to create a range of forms to show their ideas. Though these resources may seem natural for small children, the same kind of setting is needed if we are to encourage visually-based work in adults!

- Student-created works can adorn an environment more productively than materials designed just to "look good." Such displays provide students with a sense of belonging and a connection to their own endeavors. They also encourage students to learn from the work of others, helping them acknowledge the existence of multiple "good approaches" to the same problem. The typical "hand your work in secretly to the teacher to get your grade" model of learning is put to shame by such a highly visible student-involved approach to learning.

Biography

Alison Quoyeser received her BA degree in Fine Arts and Psychology from Smith College in 1973 and her Masters of Architecture from MIT in 1978. She maintained parallel career tracks in both architecture and psychology in the 1970's and 80's, and was awarded a Chance-Flanders Fellowship from UC Berkeley for a scholarship in teaching in 1987. She has taught second grade for six years.

---

66 The blackboard at the front of the room I use for a lot of purposes, such as every day I have the schedule for the day posted...Other times we use the blackboard for children to come up and demonstrate something they know...they love to do that." -Alison Quoyeser 123

66 It's a wonderful thing to see how a child who may be feeling very tentative about their performance, to see how they come to life and look stronger just because I've put their picture up on the wall." -Alison Quoyeser 124

Public Spaces:
Inviting Visitors to Learn

There are many public spaces that have been intentionally designed to present ideas. They need to provide readily accessible materials to visitors in a comfortable manner.

It's not surprising that mechanical designers like Denny and teachers like Alison give serious consideration to the visual aspects of their environments. Their jobs are directly tied to the physical objects in their worlds and to how they perceive them. Denny's primary task is to invent new objects and to explain these to colleagues and clients. Alison's job is to work with young children who are immersed in their own sensory worlds and to encourage reflection based on concrete experiences.

Visualization and the Sciences
But what about physics? How does the analytic investigation of science connect to visualization, or the visual representation of ideas?

And how does one incorporate these processes and experiences in a public space like a museum, where the typical visitor is unlikely to be a trained scientist?

Rob Semper is Associate Executive Director at the Exploratorium, an interactive science museum in San Francisco. His task is to design exhibits that illustrate scientific phenomena, exhibits that can form a basis for discussing and exploring important ideas.

This perspective builds upon a long tradition of scientific reports that discuss the critical role of imagery in motivating and describing scientific breakthroughs. Einstein's work abounds with reports of sensory intuitions and the importance of imagination in

66 The exhibits actually serve as a nice prop for teaching, and a lot of the learning that happens around the exhibits is very social... By having these stimulating tools or drawings or maps, we actually help trigger learning in a better way." -Rob Semper

developing new understandings. Kuhle reported that he discovered the structure of benzene in a dream that showed interconnected snakes revolving. Watson and Crick built many concrete models of the structure of DNA as they uncovered the basic structure of this amazing molecule. The visualization of mathematical and scientific data is currently a major focus in the computer sciences, as large amounts of data sets become accessible to human interpretation in visual ways.

The notion that perception is important to scientific understanding is at the core of the Exploratorium. Frank Oppenheimer established this innovative center to give the public the opportunity to observe and interact with physical phenomena. From these direct experiences, he reasoned, untrained people could then derive the general scientific principles on which they were based.

Oppenheimer believed that this process of observation, interaction and comprehension was the foundation of most classical scientific investigation, and should also serve as the underpinning of science education. Unusual as this concept still is to both museums and scientific schooling, the Exploratorium is thriving and its approach has influenced many institutions throughout the world.

Interactive Exhibits

Rob works with others at the Exploratorium to motivate visual thinking in this tradition, to produce exhibits that concretize natural phenomena and motivate important scientific concepts.

The exhibits themselves have then become Rob's environment, and the environment that serves the curiosity of his visiting public. Although they may be larger in scale and more public, the exhibits correspond to the personal and tutorial objects found in Denny's and Alison's environments. They are resources that illustrate and inspire and inform.

The heart of this public environment is the Exploratorium's prototyping shop. Situated near the entrance of the museum, it is a large, noisy, and highly visible space where visitors can watch exhibits being constructed. Over it hangs a sign that says, "Here is being created the Exploratorium, a community museum dedicated to awareness." In the prototyping shop are all the tools and materials needed to create exhibits, including table saws, lathes, molds, welding equipment, lumber, metal, and sheet metal. Here designers and technicians "work hammer and tongs" to bring new exhibits to the floor.

---

66 When I was studying physics in graduate school, I began to notice that I was building up this repertoire of little tiny films in my head, little dynamic visual images...if I wanted to figure out some things in quantum mechanics, I'd build a little mental model that was quite visual."
-Rob Semper 201 Overviews

Tour the Exploratorium prototyping shop on the *CD-ROM*. 223

The prototyping shop emphasizes the timeliness and the vivacity of the museum. The Exploratorium is not a warehouse or showroom for musty historical works. It is an institution actively involved in discovering new ways to explain substantive ideas.

One crucial advantage of the prototyping shop is that it offers Rob and his colleagues a place to practice their crafts of exhibit design in context. They are surrounded not only by the visitors who will use their exhibits, but also by the setting in which their work will be shown. This may seem obvious, but it's important to notice how often in our daily lives we are removed from the real purpose and intent of our work. We frequently create product designs in a lab, or product brochures on a computer screen, far removed from the consumer who will eventually use and experience them.

The Exploratorium's exhibit floor is a cavernous world where islands of hands-on displays are surrounded by a milling sea of visitors. Each exhibit includes a brief set of instructions, telling you how to use it, what to expect, and which principles are being demonstrated. Although the exhibits each represent a single basic phenomenon, they are thematically grouped to reinforce and extend the user's experience of related principles.

Rob describes a number of the museum's exhibits on the *CD-ROM*. These descriptions don't provide a full sense of the phenomena, but they do offer a sense of the Exploratorium's exhibit strategy.

In the Resonant Ring exhibit, for example, rings of different sizes have been built to demonstrate that all objects have a resonant frequency at which they vibrate. One factor that affects this frequency is the object's size. The exhibit's instructions articulate this basic principle, and tell the user to change the frequency of a tone and then notice how differently-sized rings vibrate.

Once users have a basic appreciation of this phenomenon they can engage in many follow-up conversations, such as: "How is frequency information carried on air waves? Are there other media that carry frequency information? What other attributes of objects determine resonant frequencies? When is it important to consider the resonant frequency of an object? Why don't I notice objects resonating all the time?"

The exhibit deliberately doesn't answer all these questions. It is designed to initiate these conversations and awaken curiosities. It is a concrete visual representation of a phenomenon that can serve as a conversational prop for students and teachers. More subtly, it is an example that provides people with an intuitive sense about resonance that they can use later to interpret other relevant lectures and demonstrations.

66 I sometimes stand and watch the waves come in through the Golden Gate Bridge. A lot of people stand there and see water and beautiful waves. What I see is a physics experiment in progress." -Rob Semper

Your Visual Workspace

There are a number of things we can learn from Rob and the Exploratorium about visualization in public spaces.

- Complex ideas can be represented in concrete form. You can create a number of "idea objects" for others to explore so that they can begin to understand a new concept. These objects can encourage interaction, letting the viewer build personal intuitions about the idea displayed. This approach is effective not only in museum settings, but also in other communications-oriented areas, such as schools, marketing and advertising agencies, and design firms.

- There are advantages to actually showing how an idea or object is created, instead of hiding the often messy details of the creation process. Such openness can give your audience a better understanding of the concept you are trying to display. The prototyping shop at the Exploratorium certainly serves this purpose. Other places may accomplish this in very different ways.

- Many phenomena are invisible to the naked eye. For example, ideas about cultural or social structures, and the processes of economics, or physics, or cooking cannot be seen. If you can find a way to make concepts visible and explicit, you can encourage your audience's interest and interaction.

- Being able to create a product in context can be very effective. Direct experiential contact with the intended display area and the target audience can be instrumental in shaping your designs and ideas.

Biography

Rob Semper earned a Ph.D. in Physics from Johns Hopkins University in 1973. He taught and researched solid-state and nuclear physics at Johns Hopkins, St. Olaf College in Northfield, Minnesota, Lawrence Berkeley Laboratory in California, and UC San Francisco.

Rob joined the Exploratorium museum in 1977 to help train college teachers in interactive exhibit development.

Rob is now Executive Associate Director at the Exploratorium and its Director of the Center for Media and Communication.

Consider the differences between the Exploratorium and the classroom in supporting visual learning. Record your thoughts on these comparisons and contrasts in your *Sketchbook*.

Studios: Organizing Tools for Making Things

Frank Wiley is extremely skilled in the use of digital media, employing a range of computer-based tools to create multimedia software products.

His environment combines traditional studio elements with electronic tools. Although Frank surrounds himself with high technology objects, his environment still feels like a traditional studio. It is a place that provides him with stimulation, materials, and a nurturing ambience. Frank's "electronic studio" environment supports his work and that of his collaborators.

Frank has equipped his office with a range of electronic tools that allow him to transfer conventional materials "from the world" into the computer domain. He has set up a number of digitizing stations to accomplish this. At one desk he has a computer dedicated to creating digital sound and converting other formats to digital representations. Another station is equipped to digitize video, to take materials compiled in analog video and transfer them into the computer

where they can be edited, composed and compressed. He also has a scanner which allows him to scan print documents and convert them for use with his computer.

These tools allow visuals to be equal partners with text in presentations; because everything is in digital format, different kinds of things can be combined in ways that have heretofore been impossible. It takes images off the walls and out of drawers and boxes, and makes them easily accessible. It allows imagery to be used facilely in communication contexts, moving it from the status of interesting artifact.

66 Technically, we're pretty well equipped. We can do just about anything we can imagine. People just sort of drift in and out, doing different things. It's a big pattern with me, to have a space where people can come and collaborate. So that's what this is – a central meeting place." -Frank Wiley

One goal of all these tools is to create a set of multimedia presentations, many of which will be delivered as CD-ROM products. Each of these tools is critical to the production phases of multimedia development, enabling Frank and other designers to combine information elements quickly and directly, and to develop computer programs that tie them all together. These tools also allow the design team to quickly modify visual materials so as to fluidly consider alternative approaches to a task. They are also useful in gathering, viewing and presenting ranges of visual resources which are relevant to a project. In a very compact physical space one can move easily between visual and auditory displays, still and moving graphics, first versions and final versions.

In addition, these electronic tools enhance Frank's visualization and communication capabilities. One of his major objectives is to construct electronic prototypes, to visualize new products. These prototypes are crucial for developing the content and "feel" of a product. These prototypes are also important for communicating ideas and the progress of projects to clients and to colleagues.

Like Denny, Frank takes great advantage of physical objects for inspiration. Frank's environment contains posters, paintings, and other objects that have personal meaning to him. A uniquely personal part of his studios are the sets of model airplanes that he has constructed. The books in his studio help motivate discussions and spark core ideas that he and his team can then represent in their multimedia products. He also has a supply of videotapes and audio compact discs through which to browse. On occasion, Frank's objects let him deliberately focus his attention on things other than the task at hand in order to get his work done. He says,

"A lot of the objects here are musing tools. When I say musing tools, there are all those times when you're facing what I call 'the white wall.' It's just not coming, you're staring at this blank sheet. I just go look at pictures, grab some books, and start leafing through and start thinking about things. I find what's most useful for me is to NOT think about what I'm trying to think about." **101 Overviews**

Consider how Frank's electronic environment compares to environments where tools and products are more mechanical. What are the advantages and disadvantages of current electronically-based habitats? Sketch your ideal electronic workspace in your *Sketchbook*.

Rob's prototypes are made of levers and gears. Frank's are computer programs.

Frank has carefully engineered his environment to bring new things to himself and his team, and to keep it from the isolation to which electronically-based habitats can fall prey. Fortunately, as computer on-line resources become more media-intensive and media-friendly (e.g., higher resolution images, quicker processing times), they will offer extended sets of references, and multimedia producers can expand their working environments.

As the world takes further advantage of the digital revolution, electronic studios such as Frank's will become easier to establish. The shining promise is that each of us will have access to these tools and be able to create media-rich formats for expressing our ideas in a conversational rather than professional manner. Then our language will truly be a "visual" one, where participants "speak" as well as "listen."

And of course, when that happens, we will all need to have well-developed visual abilities to make our content as rich as the tools we use!

Your Visual Workspace

There are a number of environmental concepts that are highlighted in our visit with Frank. These include the following:

• General reference materials can help motivate new ideas and initiate discussions. These materials may include both printed and sensory materials, such as audio cassettes, videotapes and CD-ROMs. As new materials go "on-line," there will be an increasing number of visual and textual reference resources available. This can "bring the world" to you, expanding your primary environments.

• Although they allow us to store and quickly access large quantities of material, computer environments hide things from general view. It is important therefore to find ways to make these computer-based environments more visible in the real world.

• Well-organized tools can facilitate work. Being able to quickly and easily perform an action means that you will be inspired to try it out.

Both Denny and Rob's "products" are visible as soon as you enter their environments. Frank's products are invisible in his workspace, becoming concrete only as the computer brings them forth.

Biography

Frank Wiley is the owner of Media Culture, an interactive multimedia development company that he founded in 1989. Media Culture specializes in creating non-violent games and curriculum-based software. It draws on the talents of a diverse group of multimedia programmers, designers, writers, and educational consultants, all of whom are focused on developing and promoting high-quality interactive learning. Prior to founding Media Culture, Frank did design work at the Apple Computer Multimedia Lab.

66 Sometimes I become terrified of my own isolation because I've made this such a comfortable space. I don't have to go anywhere to get anything done." - Frank Wiley

Environmental Opportunities

In our four visits we have seen a range of environments, with many different kinds of uses. They are not particularly flashy or unusual environments, and the people in them are each distinct and individual.

All of these environments contain commonalities. Each of these environments is fundamentally structured to create tangible, visual representations. Each contains a wealth of objects used to inspire and extend ideas. And each offers common sense guidelines for creating spaces that link the abstract and the concrete, the idea and the object.

As you work to develop your visual abilities and those of your colleagues and students, you will need such an environment. You will need a place where you can write on walls, hang up your work, spread things out. You will need to feel comfortable and inspired. You will need the tools and objects that can help you explain what you are doing.

This chapter can help you to structure such environments for yourself, and convince you that even normal workspaces can support your thinking process if you will pay a small amount of attention to their design.

It's interesting to note that most environments have not taken these issues into consideration. High school classrooms and college lecture halls are often completely inappropriate for visual learning and working. Museums where you can't touch things are not oriented in this way. Offices that are sterile are not very helpful. Workspaces without access to tools for

Use the objects around you to inspire your work.

making things are of little use. Environments without sensory richness, without color or form or inspiration, are not conducive to visual approaches.

So be aware of your physical environment and the options you have in creating it. Think also of your social environment and how you can change its forms and values. Learn to appreciate the nature of the visual culture and use it for your own gain. This will be the topic of our next chapter. Roll the *VizAbility* cube and consider the culture of visualization.

Product design. Learning. Exhibit design. Multimedia design.

Offices. Classrooms. Museums. Studios.

Objects for inspiration. Objects for explanation. Places for conversations. Places for work. Places for others. Places for oneself.

Explaining. Showing. Telling. Comfort. Stimulation. Organization. Places for Spontaneity. Silly. Serious. Beautiful. Functional. Inviting.

Make your environment personally meaningful.

Give yourself space to try new things.

Culture

There is a community of people who—maybe even without realizing it—have entered into a world of sketches, prototypes, whiteboards, idea logs, and the other visual forms.

This world invites us to express ourselves visually...It invites us to sketch even our most abstract thoughts...to be self critical, as well as to artfully engage our colleagues in critiques...

Become familiar with the productive cycle of sketching, prototyping, and critiquing to bring your ideas from fragments to substance. **201 Introduction**

Introduction

The visual thinking culture uses methodologies that are new to many of us.

While the majority of us are trained, at least a little bit, in the problem-solving methods of the scientific culture – hypothesis testing, theory development, data analysis, and prediction – very few of us have experienced (much less received training in) the methods that are embedded in the culture of visualization.

It is clearly time for this to change. We now have powerful techniques for representation, including computer-based tools, and they are getting better everyday. This is fortunate, since our lives contain increasingly complex problems, many of which require an understanding of dynamics and interactions that are difficult to imagine, understand, or discuss without visual representations.

In this chapter we will touch on some of the primary techniques and tools used in the visual thinking culture. For this culture, like any other, has established a set of expectations and methods for daily activities. It has adopted general rules "for doing business" and for assessing progress.

Key to the visual culture which is addressed in *VizAbility* is the premise that one does not know everything. In contrast to many professional arenas where the precept is that experts know everything to even participate, the essence of the visual culture is in not knowing.

One does not know what a new book or a new house or a new product will look like or feel like before it is invented or produced. One does not know if a party will be successful or a new computer system design will be effective.

211 Sketching
221 Prototyping
231 Critiquing
241 Readiness

  66 This curriculum has been silent for too long. It is time to give it voice." -Ed Murphy, President, PWS Publishing

Indeed, as one moves along in the process of addressing these tasks, more and more is known. And in fact the entire cultural milieu, as you will find in this chapter, is addressed towards making representations that do let you know what things are going to be like before they are completed. And yet when you first take on a design task, you don't know its solution, and there is no reason you should.

In addition to not knowing, the visual culture also has quite an explicit acknowledgement of the importance of "muddling through." In the early stages of a design process it is quite acceptable, and oftentimes very desirable, to be involved in very divergent approaches to the same problem. It seems important that participants get immersed in the problem or opportunity, and get a "feel" for it that can guide their progress in almost an unconscious way.

And yet there is also quite an explicit understanding of the importance of the appropriate times to be vague and multidimensional and the times to be crisp and direct. Moreover there is a clear requirement that one should try to be quite articulate about describing where one is in an activity, in many cases describing quite thoroughly what is not known and where the difficulties are.

In order to accomplish these objectives, the visual culture has developed a range of different communications methods, each appropriate to different stages of a design process. Methods of diagramming and drawing become key to these communications, extending more traditional verbal and formal representations. They let participants share their imaginations with each other, as well as to be explicit about the opportunities they see. The representations of projects are critical as information is exchanged and congealed amongst people with differing perspectives and often different tasks.

Read about this culture in this chapter and consider how the basic methods apply to your own life and work. Ready yourself to consider seeing, drawing, diagramming, and imagining activities as the key elements within a visual culture, activities that can help you think and to be productive in the everyday contexts of going about your business.

66 To me, visual thinking is really about spatial thinking. That's why I gesture so much. Ideas take place in space. So when I was a kid, I played with wooden blocks and made shapes, and it was a very natural thing to do." -Scott Kim 201 Overviews

66 It's definitely a culture. But it's more than a culture of visualizers. It's really the culture of creators, of people who are inventing, who are willing to live on the frontiers. Visualizers, I think, are one of the vibrant growth edges of this whole information world." -David Sibbet 201 Overviews

The ARC Cycle

The visual culture rests upon a principle of cyclical iteration: Act, Reflect, and Change, also known as "ARC."

The ARC cycle is a central methodology for the visual culture and contains three basic steps:

1. Act

Make your idea or problem explicit. Find a representation of your problem "in the world" so that you have a concrete and public metaphor around which you and others can work.

2. Reflect

Reflect on this concrete representation, by yourself and/or with others. Now that you can "see the situation," it is possible to analyze it and to formulate approaches to continue your development.

3. Change

After reflection, determine alternate actions that can change your representation to something more positive. Choose among these alternatives, creating a new action that will initiate a new cycle.

A typical cycle might go like this:

An idea is visualized through informal sketches (A). These are discussed with a small group (R). It is decided to move ahead (C). Potential solutions are prototyped into a three-dimensional model (A). These are reflected upon, critiqued, and modified. This leads to more sketching and more prototyping, which in turn leads to more appraisal and refinement.

Theoretically, the process can go on ad infinitum, though in practice one usually tries to create a timely final product that reflects constraints in time, budget, and sanity.

A design methodology encourages approximations and experimentation.

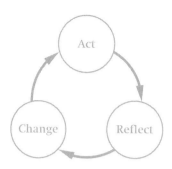

The heart of this process is an iterative cycle, one that tries to move closer and closer to a solution by a succession of approximations.

Each approximation is an action. After reflection, one modifies the approach and describes the next approximation attempt. The analysis of each approximation in the cycle provides feedback to the entire process, allowing one to benefit from earlier experiences. This feedback, key to all cybernetic analyses, then helps in the articulation of next attempts.

Literature on problem solving, especially literature concerned with the different fields of design and "right brain" thinking, has often acknowledged this cycle. Robert McKim describes it as the ETC cycle (Express, Test, Cycle) in *Experiences in Visual Thinking*. Lawrence Halprin calls a similar approach the RSVP Cycle (Resources, Scores, Valuation, Performance) in his book of the same name. James Adams addresses this same kind of iterative problem-solving in *Conceptual Blockbusting*, and Edward De Bono describes it in *Lateral Thinking*.

ARC: A Model for Learning and Invention

Although the ARC cycle is not a particularly mysterious or newly-discovered notion, it has not entered into our mainstream culture. This is partly because we consider the concrete representation of ideas and problems to be a primitive behavior, one that is appropriate to children or beginners in a field, but one that should be replaced as we learn to do more of our work "in our heads."

Reflection provides a critical feedback loop in the process of invention.

66 Prototypes allow you to explore your materials and refine your ideas. They are concrete enough for others to share and experience, yet tentative enough to advance the design process instead of freezing it." 221

We continue to naively assert that people should solve problems directly, and "get it right the first time." We often undervalue concrete representations as inappropriate and useless in addressing important abstract issues.

We also seem to think that intelligent people arrive at brilliant answers and analyses of problems very quickly, as if they simply retrieved "the correct answer" from some pre-stored set of solutions.

We forget to acknowledge the struggles and multiple approaches that are part of ingenious work. We forget all the false starts, partial solutions, and dead ends that most of us experience. Perhaps it is psychologically healing to forget these setbacks, so that we can find the daring to take on new things, again and again. Yet ignoring this iterative and developmental approach to defining and solving important problems is short-sighted, especially in learning situations. A learning theory that does not stress iteration and concrete representations might be adequate for solving some "school puzzles,"– e.g., activities that can be resolved in moments, or a few hours – but it is woefully inadequate when addressing how issues are solved in real – world contexts, such as our professional or everyday lives.

ARC: A Cycle of Worldly Actions and Mental Reflections

The visual thinking culture thrives in the alternation of action and reflection, of concrete representations and reflective considerations. In many situations it is invaluable for "breaking open" a problem or for clarifying a solution. ARC cycles are particularly useful during the early phases of a project, when they can define approaches that may still be somewhat abstract. They are also effective in collaborative situations, where an evolving idea can be shared and commented upon by a large number of knowledgeable individuals.

Though very effective, it must be acknowledged that the ARC cycle is full of tensions. Primary amongst these is the requirement to Act. For when one takes action, the failings as well as the successes of an approach to a problem are immediately obvious to many, including oneself. If one keeps one's ideas to oneself they can seem exciting and subtle and innovative. Yet in the light of the day many seemingly magnificent ideas can appear disappointing and trivial, even to their author.

Of course this is also the strength of this cycle. Once a proposal is made explicit then it can improve, gathering momentum in its own right. Also as proposals are explicit they won't get lost in the deep crevices of one's own guilt, coming under more and more pressure to appear in the world fully formed.

Inspiration lies at the intersection of bold action and quiet reflection.

Quite obviously there are also great tensions in Reflections. At first glance reflecting on a new idea can be very calm and settling. And indeed when one first gazes upon a new proposal one has created there is often a great deal of satisfaction and calm. Yet after this first glance, it is time to roll up your sleeves and get some work done. Its time to take whatever you have produced, no matter how good or bad, and try to make it better. This means looking hard at weaknesses as well as strengths, and considering other people's points of view.

In this one must then consider possibilities of both major and minor Changes. And in doing so it becomes important to figure out just how to decide amongst all the possibilities, and which risks in future development are worth assuming. Interestingly it is in the Change cycle where most politics arise, as one must realistically confront the context of a project. One must look at the demands on a project, and the losses and gains which are possible. One must consider economic and social and contractual and personal issues, and then decide how to move into a next explicit Action phase.

Many people love the tensions of this ARC cycle. They enjoy the "creative tension" and thrive on "living on the edge." Others hate these tensions and avoid them. You need to participate in this cycle and decide for yourself just how to handle these tensions, and which are important in your own work.

In the following sections two very common actions – sketching and prototyping – will be described, as will a prevalent mode of reflection – critiquing. This chapter will conclude with a consideration of change and a brief description of the states of readiness that can encourage productive work within an ARC cycle.

Reflection

Once one adopts the ARC cycle into one's work, the fundamental usefulness of this approach becomes obvious. Take a moment to consider your own approach to problems. Think of two or three issues you have confronted lately, especially ones that took more than a week to solve. Did you use an ARC cycle to address them? If not, imagine how you might have approached these problems using ARC.

Think also of your training in school. How often did you use such an approach? Do you remember teachers ever handing back your work with comments on how to improve it for the next approximation? Or did they simply tell you what was wrong and give you a grade based on how wrong you were, without providing a new venue to take advantage of the feedback?

One must begin and begin and then begin again to see fresh solutions to emerging problems.

ARC Cycle Actions: Sketching

Sketching is one of the most prominent actions within the ARC Cycle. A readiness to sketch is key to the visual culture.

Typically we think of sketching as an artistic activity, as a style of self-expression that certain people use to show what they see, or to convey certain feelings. This kind of sketching is exciting, important, and provocative; it is, in fact, a substantial underpinning of our entire society.

However, it is usually limited to a small number of people who are considered "artistic" or "talented," and generally confined to hobbyist leisure activities or the art and design professions. It is not considered a "language of the masses," one that everyone should learn much as we all learn to read, write, and do arithmetic. Nor is it considered relevant to many professions, notably it is generally excluded from analytic fields, including scientific and mathematical fields.

Yet if we consider sketching as a general communications tool, it shifts our perspective on these stereotypical uses. If we were all expected to sketch out our ideas, communities, companies, and schools would soon find ways to establish the appropriate training. And we would begin to discover a new set of professions – where sketching is used as a critical method of communication.

This latter approach is advocated, supported, and exemplified in the visual thinking culture. A rapid "sketch response" is encouraged as a way to make ideas concrete. Everyone has "permission" to draw, as the intent is not to create beautiful renderings but to place visual elements in a context that communicates ideas.

66 To me, the picture is a dynamic thing. It's not a static thing. So the process of drawing an idea, of sketching it, adds a sense of time, of dynamism, that's really important." -Rob Semper

The materials used for sketching may be as casual and spontaneous as lipstick and a paper napkin on a lunch table, or as pre-designed and utilitarian as colored markers and a whiteboard in a conference room. The places conducive to sketching are as pervasive as the sites of thought itself – occurring anywhere and any time an individual needs to figure something out, either alone or in a group.

There are two basic contexts for sketching: public and private. We'll look at each of these in the sections that follow.

Sketching by Yourself:
From Your Mind and from the World

The private context for sketching is a generally meditative state that occurs when you are sitting alone, pulling on inspirations from the world and your own mind. You might look outside at a garden, for example, and then draw a flower as mock-up art for the cover of a report. Or you might recall an experience and sketch out a few pictures or diagrams to help remember and record that event. You can also switch between the two sources of imagery – looking at the flower (external image) and remembering an experience (internal image) – to generate a visual interpretation of your experience.

Personal sketching is typically one of three sorts:

- You can sketch in order to provide a focus for your thinking, to work out ideas, make plans, and resolve ambiguities.

- You can sketch in order to remember, to capture fleeting thoughts or memories before they fade into the recesses of your mind.

- You can sketch to explain or show your thoughts to someone else.

Personal Sketching Formats: The Idea Log

Many visual thinkers carry a sketchbook with them, so that they're always prepared if they need to sketch something out. These "idea logs" can be of any size and format, horizontal or vertical, spiral bound or glued – whatever is convenient and appealing. The label "idea log" makes it clear that this is not just a place to make pretty pictures but to work out ideas visually. Idea logs frequently contain textual notes as well as sketches, to reinforce, flesh out, or archive a particular idea or event.

> 66 There's this communication on different levels that drawing is a critical part of. The first part is communication with yourself. You have these ideas, but until you get them down on paper, they remain nebulous and unrecorded." -Denny Boyle

> 66 I find that I use idea logs for many things, whether they're little projects at home or events that I'm planning. I even keep a life idea log... Writing things down is a good habit to get into, and adding sketches or pasting things in, any information that's related." -Denny Boyle

Idea logs can act not only as archives, but as prompts for fresh approaches to a situation. Denny Boyle has kept idea logs for years, for example, and often refers back to them. He uses them to explain his current ideas and to re-examine his earlier assumptions. They are vital creative resources that he can call upon, both during the course of a project and after the project is done. **112, 114**

The *Sketchbook* can become your personal idea log. Use it for keeping notes about this *Handbook* and for doing the exercises on the *CD-ROM*. Try to sketch something that catches your eye, or to capture a fleeting idea that dances through your consciousness. Use it to reflect on visualization and visual opportunities in your spare moments, in random places.

You can also use the Notes feature on the *CD-ROM* as a text-based idea log. Or, if you have more elaborate word processing and graphics applications, you might want to create your own electronic idea log by combining text with computer-generated sketches and materials from the *CD-ROM*.

When sketching is a public experience, it can become a critical element in communicating and collaborating with others.

In this context, a sketch is seldom a static element, a picture to be judged apart from its creator. It becomes, instead, a context-sensitive object that changes its form as the ideas it represents evolve. It can serve as the record of an idea, or the core image around which ideas develop, or the beginning of a more detailed analysis. Instead of being an isolated object, it becomes the center of a communications activity.

Public Sketching Formats: Boards

Collaborative sketching can be done on all sorts of surfaces. Idea logs and restaurant placemats are likely places for two or three people to sketch together. For bigger groups, larger surfaces are required to provide easy access for both "sketchers" and "watchers" who need to rapidly alternate between these two roles.

66 There's something about the process of putting it down in a way that you can see it that allows you to speculate, to see new things, to make it more tangible. I think a lot of this is an interplay between the outer world and the inner world..."
–Rob Semper

Whiteboards and blackboards are two common surfaces, often provided in active visual environments to encourage people to sketch out their ideas together.

Blackboards

Classrooms almost always have blackboards. Typically, these are used by a single person who is trying to explain an idea, and often the activities are exclusively text-based. In these contexts, the blackboard acts as a spatial organizer for the presenter and offers a somewhat permanent record of the ideas presented (more permanent than words floating through the air).

On occasion, blackboards are simultaneously or collaboratively used by a number of different people. This can be observed when an entire class goes to the blackboard, or when a few students work out an idea with their instructor. Here again, however, text often dominates, though the media allows more visual displays. As more educators and communicators become aware of the benefits of a visual culture, images should begin to take their place on these public displays.

As many of us recall from our own school days, "going to the blackboard" can be a rather a high-anxiety experience. Some of us remember the embarrassment of failing to solve a problem at the board, or of having to write on the blackboard as a disciplinary action. Yet this is not necessary and can change.

Whiteboards

"A whiteboard is an important element. It puts ideas up in lights, it projects them... You can leave your sketch up there and it sort of stays as a symbol of things that have happened. It's very temporary because you can get rid of it quickly... It doesn't encourage very accurate drawing at all...it's a quick recording of gross concepts." Denny Boyle

Whiteboards are relatively new public displays of text and graphics. They provide a surface that is slightly more conducive to detailed drawing and composition than blackboards. Colored pens can give a richness to these displays, and allow for more detailed, multi-level sketching. Yet the temporary nature of these displays makes it explicit that this is still a place to work out ideas, rather than impress colleagues with a lot of fancy graphics.

How many times in the last week have you sketched together with someone else? How many of your environments have accessible spaces for pictorial conversations?

❝ The visual culture encourages people to sketch out their ideas wherever they are, on whatever is at hand: a napkin in a restaurant, the sand on a beach, or a board in a meeting room. They sketch to capture their ideas and to show them to others for discussion and elaboration." 211

Whiteboards are common in research environments, in industry and in academia. They seem incredibly cordial to places where idea invention is valued and encouraged.

These boards come in a variety of sizes. Small whiteboards are found near personal workspaces, whereas large whiteboards are frequently available in conference rooms. Floor-to-ceiling whiteboard "walls" adorn many corridors and research spaces. **642**

Collaborative sketching is a frequent activity on whiteboards. One person starts a sketch and others join in as the conversation and drawing evolve. Sometimes the interaction is centered around short text phrases. Often, arrows, lines, and circles are drawn to connect these phrases as the dialogue progresses. More formal diagramming techniques may be incorporated. Drawings of real world objects are often included in displays.

Group Graphics

On occasion this kind of visual interaction is formalized into a group diagramming session. People come together to brainstorm, to create as many ideas about a topic as they can. They may alternate turns at the board or sketch and write simultaneously. **513**

Sometimes the meeting is facilitated by a leader who coordinates the graphical representations. David Sibbet, an innovator in this type of communication, calls these activities "group graphics."

The images used in graphical conversations are highly context-sensitive – that is, they are often unintelligible outside of the original situation. This is in contrast to more formal graphics, which are generated to be read and understood independent of their initial context.

Yet, for the participants, the graphics prove to be invaluable. They provide a focus for conversations, they record ideas that would otherwise be lost, and they create a record of the meeting. At the end of a session, the full whiteboard can offer a sense of accomplishment and, in well-managed meetings, a sense of closure.

Given experience with whiteboards, many people become quite dependent on them to do collaborative work. They find the board a natural accompaniment to productive work, and a good tool to encourage thinking.

66 If you're really interested in communicating your idea to the group, you get up there on the board and you try and give the colored markers a chance, and you work your way into somebody else's idea." -Denny Boyle

Unfortunately, many people have not had experience with these public sketching formats. Our mainstream culture has relied largely on verbal exchanges, expecting innovation from individuals working in isolation, rather than from groups generating dynamic displays.

As you develop your own visual abilities and as you come to rely on the environments that support them, you'll start to notice that most places are not set up to encourage the use of visualization. So it will be up to you to become conscious of this and change the environments you live and work in. There are many blank walls just waiting to have whiteboards installed on them!

ARC Cycle Actions:
Prototyping

Prototypes can make an idea explicit in three-dimensional and functional arenas.

The two-dimensional quality of a sketch on paper can limit one's ability to show an idea. This is particularly true when a three-dimensional object is the focus of the communication, when the topic is the design for a new toaster, or a house, or a bridge. Sketches are also limiting when the "product" is actually a dynamic or interactive process, such as multimedia CD-ROMs, urban planning models, exhibits, or physics experiments.

In these cases, prototypes are the actions at the core of the ARC cycle. Building a prototype allows you to refine ideas in the material world. Its "physical reality" lets you share your ideas with others by direct example, while its tentativeness acts as a placeholder, advancing your understanding instead of freezing it.

What exactly constitutes a prototype? The definition is actually broader than most of us think, encompassing all sorts of exploratory pre-production. The rehearsal of a play is a form of prototyping, as is putting together a new recipe, or the act of trying on different clothes before a party. In a sense we are all prototype designers.

66 One of the important features of prototyping is taking something that you imagined in your head and putting it out in the concrete world...you get to manipulate it and test it and see if it's working..." -Rob Semper

Diversity of Prototypes

Even if their content and details are different, sketches on paper naturally have quite a bit in common with one another. Prototypes, on the other hand, can be extremely diverse. They range from "quick and dirty" mock-ups (a house modeled with cardboard boxes, or an airplane made of balsa wood and glue) to very elaborate and expensive designs (an architect's model of a city center, a working representation of a pulley system). Prototypes can be large (an automobile modeled to scale) or small (a wax casting of a ring). Prototypes may be built to the scale of the final product or may be smaller or larger than the proposed result.

Prototypes are useful at different stages of a design process. Very rough models are typically constructed at the beginning of a project cycle, while more sophisticated models are created for the final stages of design. Early prototypes are usually made for the people involved in a project, while later versions are intended for client approval or customer feedback.

Yet despite their diversity, all prototypes have the same intent. They are designed to try an idea out. They represent concrete actions in the world that can be viewed and evaluated, and then modified.

Good Prototypes

The creation of a good prototype is an art. It is critical that a prototype addresses certain key issues and that it contains a level of detail that can help evaluate these issues.

A good prototype has two basic characteristics: it is appropriate to the level of a given idea, and it is designed to help distinguish the alternate possibilities under consideration. Its overall goal is to allow decisions to be made. A prototype differs from the final desired object or set of ideas in ways that are appropriate for the analysis at hand. It is not unlike the design of a good scientific experiment, where critical features must be maintained in order to allow generalizations, and where variables must be controlled in order to allow central issues to be identified.

Obviously, certain activities benefit more from prototyping than others. Yet when you realize that prototypes can be built to represent even abstract and dynamic processes, the potentialities of prototyping are compelling.

---

66 The process of doing real stuff is really what leads you to new insights about the next iteration of the design." -Rob Semper

66 These little models don't hold up really well over time, but they get the message across, especially in the first few days when you're trying to convince not only your client but yourself that you have a good idea." -Denny Boyle

The Prototyping Environment

A supportive environment, rich with the diverse tools and materials required for imaginative mock-ups of ideas, is needed in order to encourage prototyping. Special, customized tools may be required for more complex production prototypes.

Most environments are ill-equipped for prototyping since it is not considered an activity central to most concerns. However, there are a number of environments that are set up to encourage prototyping. Not surprisingly, some of these are in the design professions, particularly in areas such as product design, architecture, and packaging. Others are in scientific arenas, where trial situations are set up for analysis and feedback before larger scale methods are implemented. Business enterprises often encourage prototyping as well, gathering materials and procedures together for market research and feasibility studies.

The Machine Shop at the Exploratorium is a good example of an environment that supports prototyping. Here exhibits are continually being built and refined before being placed on public display or shipped to other museums. The basic philosophy at this science museum is that exhibits "are never really done." Instead the Exploratorium is viewed as a research environment, one that continually seeks different ways to display phenomena and explore scientific ideas.

Every exhibit at the Exploratorium is considered a prototype, even the older ones that have been copied by other museums. As soon as they are built, exhibits are used by staff designers, scientists, and visitors to the museum. The feedback derived during this process spurs further modifications and refinements.

The exhibits are always subject to progress. In theory, this means that even the most popular display will be changed if it is judged that improvement is possible. This is quite different from most situations, where a prototype is considered to be an intermediate step on the way to a final performance, product, or system.

To support this high emphasis on prototyping, the Exploratorium Machine Shop is staffed with a large number of professionals who are expert in a range of crafts, from metal work to carpentry to plastic molds. They are both producers and collaborators in exhibit design; they generate materials and advise designers

66 Prototypes don't always happen as quickly as you'd like. Sometimes you work along and get stumped by a problem... then later you get an idea of how to resurrect it...so it's not always a smooth process. It may be a process that goes through ebbs and flows of development." -Rob Semper

on the aspects of particular crafts. The workspace itself is full of materials and tools – a real tinkerer's delight! Some resources are for long-range productions, while others can be grabbed quickly to show an idea or test out a basic concept.

Most environments do not need this level of prototyping support, since they are not as focused on prototyping as a central institutional philosophy. Yet the principles of prototyping can be applied to most productive environments. It is a process that allows one to make changes without high risk or cost, since ideas can be tested before new projects are fully implemented. They are concrete enough to motivate conversation and feedback. They demonstrate explicitly how far along a project is, helping to gauge the amount of work yet to be done.

66 In building a prototype, the real secret is in figuring out what the important elements are that you really have to pay attention to – and what are the secondary things you can worry about later..." -Rob Semper

Do you use prototyping? What have you used it for in recent months? How was it effective? Ineffective? Think of some compelling uses for prototyping. Why are they compelling?

ARC Cycle Reflections: Critiquing

The primary purpose of sketching, prototyping, and other actions is to make an idea concrete enough for reflection and feedback.

Sometimes this is not obvious. One is usually so immersed in creating a prototype, and so fatigued from producing it, that criticizing it can be a very disagreeable notion.

Yet, if we don't step back a bit from our prototypes and consider the messages they convey, they become little more than procedural exercises. For at the heart of the ARC cycle is the "R," the reflection that motivates change and provides a mechanism for improvement.

Critiquing from the Inside

Personal critiques are a crucial starting point in this process. In the simplest case, these reflections are made by the author of an idea. One completes a "draft" or prototype, then steps back and looks at it – trying to decide if the concepts are coherent, the ideas solid, and the representation appropriate and informative. This reflection can be brief and intermittent, with the author producing for a few minutes, stepping back and reflecting, producing for a few minutes more, stepping back, and so on. More typically, however, it follows a pre-established schedule of milestones, where – at certain key intervals – one halts one's work and takes stock.

In order to benefit from this kind of personal critiquing, you need to master a good attitude. You shouldn't be overly critical of your work, for example, or you may stall productivity and introduce unnecessary fears and anxieties. On the other hand, you can't afford to be too complacent, either. You must consider very seriously the notion that a project can be a complete failure; you must also make an effort to see and correct any

> 66 It's the ability to rise above your normal, everyday existence and be aware of everything you touch and use, and see other people touch and use and interact with...and wonder if it couldn't be better. You're kind of a critic, constantly..."
> -Denny Boyle

deficiencies, so that the product can be successfully refined.

In many situations, the critiquing process involves a team of people who are working on a project. All the members need to take a step back to see the project's overall progress, and to reflect upon it from the "inside."

Critiquing from the Outside

Personal critiquing is often limited because the author(s) are too close to the project. Thus, as a project becomes more developed, it helps to take advantage of an outside point of view.

In the design professions, the ritual of critiquing is commonplace. It occurs as a group process in which people identify the positive and negative features of an evolving product and convey this information to the product's creator(s) in a manner that encourages improvement.

There are always two sides to a critique, those giving it and those receiving it. While the number of people on each side varies, depending on the situation, the introduction of outside feedback can generate a great deal of emotion; the experience can be highly productive or incredibly debilitating. Problems often arise because people are incredibly resistant to criticism and few people are good at offering it in a way that is helpful rather than hurtful.

Although people know intellectually that they can improve their ideas by receiving feedback, they still tend to defend their actions and territory against outside intrusions:

"Other people may need to improve, but clearly I have enough reasons for my actions, and shouldn't have to explain myself, or listen to naive evaluations!"

Yet to get the best product, one needs to know how to work collaboratively with others, to welcome their talents and to be responsive to good ideas for change. We'll explore some guidelines that can help you improve your ability to both give and receive feedback, but first let's look at two common milieus for critiquing: the classroom and the workplace.

Critiques for Teaching

One prevalent use of critique is between student and teacher. The student's job is to create something that is expressive and appropriate to a given task. The teacher's job is to acknowledge the student's effort and then provide guidance that suggests other approaches or ways to improve the product.

This process is familiar to students in the fields of art and design. Projects almost always receive "crits," incrementally as a teacher views a work in progress over a student's shoulder and more formally at different stages of the task. This approach, however, can be quite unfamiliar to students in other disciplines.

---

66 I think the most important things in critiquing student work is to look for the positives in what they've done, the strengths, and to take care of their egos. Then it's much easier to introduce the idea that maybe they could work harder on a certain part..." -Alison Quoyeser

66 The question of whether or not to mention negative aspects in the critiquing process is a huge question...because teachers are very unwilling to say anything bad about the work... I've had students come in and say... How come the teachers don't just call a spade a spade?" -Terry Irwin

A notable exception to this is the manner in which creative writing is currently being taught in certain schools. Here students show drafts of their projects to teachers and fellow students, asking for feedback as they develop and express their ideas. Especially since word processors have made these changes possible and easy, both teachers and students can engage in group critiques and participate in a genuine ARC cycle.

Surely this approach can expand to other disciplines if we will only acknowledge the importance of feedback in learning!

Professional Critiques

In the design professions and in many engineering activities, critiques are already used procedurally to enhance the quality of work. These are first done "among friends," the team that is invested in the project. Here everyone is obviously aiming for a quality product, and wishes to maintain the energy of the team. Colleagues offer the best perspectives they can on others' work, supplying feedback about the current state of the project.

Because everyone is invested in improvement, both negative and positive emotions run high. Comments are often directed at or interpreted as criticism of personal ability instead of the objective product.

Some typical comments might be:

- *"If you weren't so blind and close to the project you would see that this makes no sense!"*

- *"This is really inconsistent. Who is in charge here?"*

- *"It's too bad you couldn't afford a good artist; the ideas here have great potential but with the way in which they're presented, they make no sense."*

- *"This is competent, but completely uninspired."*

- *"I think we need to rearrange this team somehow to make this work!"*

Somehow the language of criticism gets completely entangled between creators and creations, people and their projects. Designers' expectations – both for themselves and for their colleagues – run high. As the intensity of a project cranks up, as both dollars and reputations are at stake, there can be frequent flare-ups during professional critiques.

It is often the case in professional domains that "managers and supervisors" judge the progress of individuals through their projects, but the projects themselves receive little direct attention. Typically, they are reviewed only when they are completed and, even then, rarely by the creator's peers. In some cases this may be appropriate – not all activities are best approached with an ARC cycle! Yet it's important to consider just how and when a critique fits into one's daily life and work and how it can serve as a central part of creating excellence.

> **❝** To dig in your heels and set up a confrontational dynamic kills all the creativity. All the humor leaves the room and the creativity goes south. I think it's really important to set up an environment of permission and humor (for a critique) because it can very quickly turn sour..." - Terry Irwin

A second, noteworthy aspect of professional critiques is that they are frequently based on economic issues – that is, they focus on the quality that can be achieved within economic constraints.

"We critique on two levels, actually, because we are trying to pay our rent and buy our groceries doing this. So we have to critique on an economic level as well as an aesthetic level. I hate to say this, but it's a reality that the economic critique sometimes outweighs the aesthetic." Frank Wiley **232**

The standard adage about this is, "We can give it to you cheap, we can give it to you quick, or we can give it to you right. We can sometimes do two of these, but we can never do all three!"

A critique session or some other kind of reflection experience allows all of these different issues to surface and to become explicit. Often the critique begins with a consideration of the quality of the work, but as alternative changes are considered, economic and time constraints end up affecting the final decisions.

Giving and Receiving Critiques: The Feelings and the Constraints

There is an art to giving and receiving good critiques. Some people are quite good at one or the other of these, and a few individuals are excellent at both. In giving a critique, your goal is to offer good advice in a manner that allows it to be heard.

There are a number of ways to assure that you will be heard. The most important of these is establishing a basic trust between you and the person who is receiving the critique.

Mutual Trust Is Key

In some cases, such as the relationships between a student and a teacher, trust has been established over a long period of time and a great number of interactions. Students respect the views of their teachers, and look to them for wisdom and guidance. Similarly, the teacher expects a great deal of the student and wants the student to succeed.

Many professional relationships generate the same kind of long-term trust. Professionals have a basic respect for each other and typically work in situations where everyone has a stake in the success of a project.

---

66 This idea of critique... You have to do your own critique. You have to be your own toughest critic. You have to set up processes which make you question the refinement to its finest detail." -Primo Angeli

66 Maybe it's just me, but could you explain it? I want to make sure I understand." 234

In both of these cases – educational and professional – this trust can be fragile. People can sometimes be motivated by selfishness, ambition, and power. When the relationship becomes based primarily on these rather than trust, it produces a terrible environment for effective critiques; the results are loud, noisy, and emotional, but not effective!

In many other situations, trust must be established during a particular session, as people who are unfamiliar with each other often participate together in critiques. In these instances, the surface form of a critique becomes crucial. What one says and how one says it becomes especially important.

Provide a Positive Context for Your Comments

One popular rule of thumb in giving a critique is to say two positive things about a project before saying something negative. Like all rules of thumb, this guidance is completely useless if the critique is delivered in an insensitive manner; it's not hard to tell when someone's compliments are perfunctory, setting you up for a larger fall.

In a trusting atmosphere, however, this approach can be very successful. In describing the positive elements of a project, the critic reveals their understanding of the work and acknowledges the effort involved in producing it. If well-considered, these comments can pave the way for considering improvements to the project.

Here are some examples:

- *"This is a great approach to the project. I like the emphasis on the user as the center of the design, and i think that your use of color is inspiring. Now, if you can just provide the detail to the interface to explain it, you will be in great shape."*

- *"The use of different-sized rings to show differences in resonant frequencies is a great idea. I also think the controllers have been set up in a very intuitive manner. Now, how are you going to construct a general explanation of the resonant frequency phenomenon? And how will you provide visitors with more information about this phenomenon if they get interested?"*

- *"Your poster does a good job explaining the basic phenomenon of photosynthesis for the science fair. It's clear that you understand the critical issues involved, and you've come up with a good visual layout. Now think about what you might add, or how you might change your layout to get the attention of the viewer. You are interested in this phenomenon, but why should anyone else be interested?"*

66 For some types of products, it's possible to set up a critique where the person doing the critique doesn't even know they're doing it. You can just put something in front of people and watch them use it. That's what user testing is all about."
-Scott Kim

Be Sincere

Another primary rule of thumb in critiquing is to be sincere. Even when a project isn't very good, it can usually be improved with some well-considered guidance. The key is to say what you really think – both good and bad – and to do this in a direct manner that commands respect. If you can be honest yet courteous, this approach can be extremely effective. In reality, most of us cannot pull this off very well; many situations do not inspire sincere sweetness, and some of us are simply not very thoughtful people. Yet, even in these cases, it is important to realize that the goal of the critique is to be heard. If one is too gruff or mean or dishonest, there is little chance that your perspective will be included or that your recommended changes for the project will be implemented.

It is also important to be clear about what you "know for sure" versus what is simply your best guess at the moment. Often, for example, a teacher sees quite clearly that something is wrong in a project and that it needs to be changed. Or an experienced professional recognizes a pattern that he has experienced before and that must be avoided. In these cases, it is crucial to emphasize the need for correction in order to achieve an effective product.

In other situations, the person giving a critique is unfamiliar with a specific situation and can only offer his or her best guidance, based on general experience. It's important to let the person receiving the critique know which context you are working within, so they can distinguish between suggestions and imperatives.

One Small Example

The *CD-ROM* contains an example of a generous and thoughtful student-teacher critique between Alison Quoyeser – the second grade teacher we interviewed – and a nine-year-old, my daughter Erika.

The context for this critique is a student book report on a famous artist. Erika chose Georgia O'Keeffe. The assignment was to reproduce one of the artist's works and to comment on the experience. (Earlier book report assignments included making mobiles, developing brochures, and creating advertisements, all very visual and project-based tasks that explored a range of representational forms and possibilities.)

Erika describes this project, which she did the year before as follows:

"First I did the background, then I did the puzzle pieces, but then by accident I spilled some water on it, so I did the whole thing; I thought it looked nice the way it faded away in the frame...

> 66 As I get to know the class better, they can take more criticism from me because they trust me. Even if I say 'You need to do this piece of work over again,' it's not going to be a crushing blow because they know I've enjoyed their work in the past...they've just flubbed up this time..." -Alison Quoyeser

I used watercolors for this. First I used pencil and then I did it over in black crayon...then I painted it, I mixed some colors, just some...that's the yellow and the red.

It doesn't really look like a flower but that's what it's supposed to be. Georgia O'Keeffe had a pack of imagination in her head to do this." **232**

As you listen to Alison critiquing this project there are a number of elements you should notice. For one, she continually refers to the process Erika used in creating the project, keeping the dialogue focused on the making of the work rather than the final piece as a completed object. Some examples are:

- *"I would have liked to have watched you paint this. Did it go really quickly or did you take your time?"*

- *"Did you go back afterwards and add these lines?"*

- *"Was this part of the actual flower?"*

She also asks Erika a range of questions, finding out how Erika views the work in order to frame her own comments on it. In doing so she continually keeps the focus on the project and not the basic capabilities of the "artist."

"I like...the stamens of the flower. If you were going to do such a painting again, would you do anything different?"

She includes humor and good will in her remarks, emphasizing that the work is of merit even as there are a number of things "to try next time."

Much of the "niceness" of this critique is more appropriate to a young child than an adult. Yet the elements of this process – one that went on intensively for more than 20 minutes – illustrate a viable way of delivering an effective critique. If you acknowledge the fundamental vulnerability of all "inventors" who reveal themselves in their work, you'll find many elements of this interaction that can be relevant to an adult critique.

Keep an Open Mind

Giving critiques takes an even-handed approach, but receiving a critique requires a balanced perspective as well. You never know exactly what kind of critique you'll receive, and more often than not, you won't get the one you expected. Indeed, others will probably notice the weaknesses you've already acknowledged in a project, but they will also bring perspectives that you hadn't even imagined.

66 We wouldn't be human if we didn't have a little bit of discomfort associated with the critiquing process."
-Terry Irwin 232

"The basic premise of critique is, I've taken this idea as far as I can, or at least as far as I can for right now, and if I listen to other people and take in what they say, it can be better."
Scott Kim **232**

You need to keep an open and receptive mind in order to take advantage of these perspectives, and you need to maintain a non-defensive attitude so your own opinions don't get in the way of listening to what others have to say. (When it comes time to consider how the project should be changed, there will be plenty of opportunity to offer your own perspective on the problem; initially, however, you are in a critique to listen.)

An Experiential Example

All of this may sound quite reasonable in the abstract, but for it to become meaningful, you need some direct experience with critiques.

To get started you might try the CRITIQUE GAME on the CD-ROM. Here you will find a broad assortment of silly and profound critiques through which you can browse. To make this game significant, you should imagine that you are in a critiquing session, showing a new project that is important to you. You have just finished making a careful presentation of your work, and now your colleagues are offering you their responses. Some of these make you feel good:

"Now it's working. Can you see how you've finally solved the problem?"

Others make you want to hear more so that you can revise the project appropriately:

"I think I see where you're going. Now if you can just put a slightly different spin on it, you'll be right on target."

Still others make you want to leave the room or shrink into an invisible ball:

"The stuff you've been doing is fantastic. THIS is a disaster!"

Try a number of these critiques. Some will be irrelevant to your situation as they have been designed to be very general. Others will seem silly, and many are. Others will probably ring true to you.

As you listen, decide if you want to hear more from this person about your project. Do you feel congenial toward their remarks? Do their remarks convey a sense of respect? An agreement in basic principle? Or do you feel misunderstood, resentful, indifferent?

Although this game is somewhat fanciful, it can still help you understand your basic preferences for critique styles and your lack of receptiveness to certain approaches. You can also analyze the elements in what you considered the most effective critiques, and use them in developing your own technique. These reflections can guide your own behavior in both giving and receiving critiques.

---

66 The critique process puts a concept out into the world for analysis and feedback. In a collaborative atmosphere, this can lead to a fresh flow of ideas, or result in the refinement of existing concepts. An effective critique should always empower the artist and encourage inventiveness." 231

66 I can't articulate what it is, but something's just not working in it yet."

ARC Cycle Changes: Opportunities and Realities

We have discussed both the Actions in a particular ARC cycle – sketching and prototyping – and the Reflection – critiquing. Following these, Change becomes the dominant element in the cycle.

It is the consideration of change that drives each new set of actions and reflections. The kinds of changes that occur in a piece of work are extremely context-sensitive, to the task at hand and to the domain of the work. Yet simply stated, there are four basic kinds of changes:

- Changes that move a satisfactory product along.
- Changes that modify a product in small ways.
- Changes that drastically shift a product's direction, revamping "what might be" by looking at "what is."
- Changes that dictate abandoning the product altogether.

Move onto the Next Phase or Declare a Project Done

One very fulfilling response to a critique is the judgment that a project is wonderful and that it is finished.

More commonly, however, a project is judged to have met some, but not all, of its milestones, and moves on to the next phase of development. The question becomes "what's next?" instead of "is this any good?"

Often a critique will identify one direction from a set of three possibilities, for example, and the actions initiated by this will move the project on toward these developments. Or a critique will suggest a new mechanism that solves a particular anticipated problem, steering the project in a direction that was once considered impractical.

66 I'd say that design has possibilities...(but) we're not going to do it." 234

The discussions that accompany these changes are often charged with emotions, and usually involve alternative perspectives and competing values. They basically boil down to two considerations, "what is good enough," vs. "what is possible." In effective invention processes, these active discussions give way to solid decisions that determine the next steps. In ineffective processes, indecisions take up the time and effort that could be directed toward moving a project forward. Building mechanisms that allow these decisions to be made succinctly and directly is a critical part of setting up effective ARC cycles.

Make Small Modifications

Critiques often identify specific elements of a project that "aren't working." By discussing changes, participants try to invent procedures that will satisfy all the goals and constraints (economic, aesthetic, time) of the project.

As these elements are identified and changes are designed, the major task becomes the assignment of priorities. Which changes are the most important? Are there any expected interactions among these? Is there a newly formulated statement of goals against which the next set of actions might be judged?

It is important to evaluate what constitutes a "small change," and to consider what it will take to make this change before it is attempted. It is important to distinguish between small changes that take two hours versus those that will take two weeks.

Change Directions

A third kind of change is to drastically shift the direction of a project. Such a judgment is often made at the beginning of a project, and is typically avoided later on, since radical shifts become increasingly devastating as a project progresses.

The judgment to move in a new direction is often considered a positive step and not a failure. In many cases prototypes, sketches, or other representations of a solution reveal new information about the nature of the initial problem. In this sense they aid in "problem finding," which can be extremely helpful in accomplishing objectives.

Abandon the Effort

In extreme circumstances, a critique may result in the judgment to terminate a project. Considered in a broad context, this judgment can be productive, as it removes resources from an unsuccessful effort. Usually, however, there are large costs (to at least someone) when such a judgment is made. In student-teacher situations the cost is often the failure of a student and an accompanying negative impact on the student's confidence. In professional situations it is often associated with loss of revenues and maybe the loss of a contract or job.

If an ARC cycle is to have integrity, the choice of abandonment must always be an option. For unless one is ready to judge that a project is not working, it is impossible to consider changes that might make it better.

66 ... Maybe we should put that in – but if we do, what are we going to do with this information?" 234

Readiness: A Backdrop for Productive Work

The inherent uncertainties involved in a change-centered model such as an ARC cycle can generate a large set of demands and tensions, which individuals must address to maintain their readiness.

Because of this, individuals working with this cycle must pay extremely close attention to their own mental alertness and their ability to ready themselves for ever-changing challenges. Methods of relaxing and focusing, of listening and reflecting, can enhance most human activities. In the context of visual thinking (and rethinking), these methods can prove critical to productive work.

What Is Readiness?

A state of readiness is one that is balanced between relaxation and focus, a condition of physical and mental alertness that is unhampered by fatigue and distraction.

This seems obvious for sports and other physical activities. Athletes are encouraged to train their bodies and focus their attentions when performing physical tasks.

Yet, it is an equally important prerequisite for visual work. The process of visual thinking is composed of a series of conscious and unconscious interactions. These require flexibility of mind and the energy to represent new ideas, states of being that flourish with mental and physical preparation. One can benefit greatly from stepping back a bit, relaxing, and then "tuning in" to the work at hand.

Michel Cziksimihalyi, a professor at the University of Chicago, describes a general "flow experience" that is sought after and attained by people working in many domains. It is a state "between boredom and anxiety," a state that includes enough stimulation for

66 Preparing to imagine, invent and manifest new ideas requires a...state of readiness. Tension, fatigue and dullness hinder productive thought. By applying a few simple techniques, you can refresh your mind and body and find new energy for invention and communication." 241

creative work, yet at a comfortable enough level so that fears do not discourage productivity. Others describe this state as "being on the edge," working at a level that is challenging but not terrifying.

How Can You Achieve Readiness?

Though this "balancing act" is not specifically taught, one can take responsibility for creating a state of readiness before approaching a task. The *CD-ROM* offers three features that can assist you throughout the product: NOTES, WARMUPS, and MUSIC.

- NOTES gives you a place to reflect and archive your thoughts.

- WARMUPS contains a set of exercises that encourage you to establish and maintain productive alertness.

- MUSIC offers you a selection of different musical backgrounds to use as you engage in various visual activities.

The READINESS section on the *CD-ROM* shows you how to combine these different features. It suggests activities ranging from basic physical stretches to mental puzzles, from group activities to sketching exercises, from listening tasks to imagination challenges.

Much of the effectiveness of these resources comes from increasing your ability to pay attention to relevant elements and to ignore those that are not productive. William James noted the phenomena of "selective attention" in the early part of this century. He described what he termed the "perches and flights of consciousness" that occur in mental life and the "fringes" around conscious activity that provide context and background for this activity.

Your task is to become aware of your own consciousness, to take advantage of your unconscious in order to enhance your work, and to motivate the "perches and flights" of your consciousness so that you can get your work done!

Relax

WARMUPS contains activities designed to relax you physically and mentally. Though some of them may seem whimsical (and deliberately so), they can help you to release tension and ready yourself for any kind of inventive effort.

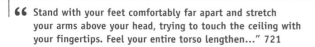

66 Stand with your feet comfortably far apart and stretch your arms above your head, trying to touch the ceiling with your fingertips. Feel your entire torso lengthen..." 721

Stretching, for example, releases tensions and toxins in the muscles, restores circulation to the body and brain cells for increased alertness. Such relaxation exercises make you feel fresh and ready to approach new tasks.

"Raise your shoulder toward your ears until you feel a slight tension in your neck and shoulders. Hold this position..." **721**

Focus

To be productive, you need to be relaxed – yet you don't want to be a "limp rag." Your mind should be alert and engaged in the task at hand.

Word games or imagination exercises can help you focus your mental abilities. Try some "free association" with the following words:

- Dragon • Paper • Desk • Dawn • Sparkle

WARMUPS provides a number of these mental gymnastics for you to try. Use these exercises selectively, to focus your attention and "clear your mind," readying yourself for new productive work.

Listen

Sounds – both musical and non-musical, such as sound effects – can also help you focus your attention upon your task. Or, if you prefer, they can take you far enough away from your task that you can gain a new perspective. In the simplest case, you might turn on some background music as you work, and experiment with how different kinds of music affect your mood and approaches.

Reflect

You can sometimes positively affect your readiness by simply taking a moment to consider what you are doing; this allows you to reflect on a particular process in order to understand it and take advantage of it. You might want to write in your *Sketchbook*, for example, or compose a diagram on a whiteboard. Or you might want to use the NOTES feature to keep notes on your progress.

Your task in all of this is to find out what works best for you. What kinds of exercises, music, and reflection help you to approach your work confidently and productively? How can you focus attention on your task and still consider alternatives? When is it time to take a new approach to a problem and how will you know it?

66 Starting from the center point on a piece of paper, draw a circular spiral. As the rings of the spiral grow, try to keep its lines parallel to the previous rings, with roughly even spacing between them..." 761

All of us are unique in our methods of doing effective work, and different situations invite different approaches. Practice a range of approaches and find what works best for you. Consider resources that can assist you in enhancing your readiness.

Joining In: Becoming Part of the Visual Culture

Cultures are made up of elaborate sets of rules and conventions that are basically invisible, whether you are totally outside a culture or completely inside it.

Attempting to describe a culture, particularly in words, is inherently a futile effort. Yet, many of you on the brink of entering this culture can, for a moment, consciously notice things about it.

You can tell that you are becoming a part of the visual thinking culture when:

• You find it hard to believe that all classrooms and conference rooms don't have whiteboards.

• You begin to sketch actively in conversations without saying "I don't know how to draw."

• You routinely stand up from your work to stretch and perform some mental exercises to ready yourself for the next push at your task.

• You interrupt your friends while you're driving to point out something you've just seen that is really neat.

• You can't function without carrying a sketchbook with you.

• You choose restaurants that have paper placemats or tablecloths (with crayons!) so you can draw while you eat.

• You look at objects, activities, and scenes and imagine "what could be" by considering "what is."

• You actually enjoy presenting an idea for criticism so that you can add to your ideas.

Of course, in order to really appreciate this culture from the inside rather than the outside, you will need to participate in it directly. To do this you will need specific skills. These skills are the focus of the next four sections.

Roll the *VizAbility* cube and begin to develop your Seeing, Drawing, Diagramming, and Imagining abilities.

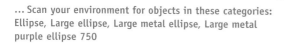

... Scan your environment for objects in these categories: Ellipse, Large ellipse, Large metal ellipse, Large metal purple ellipse 750

Seeing

Seeing is more than just looking. It involves increasing your visual flexibility...unblocking stereotypes...

refreshing your attention...noticing patterns...becoming aware of movement and space.

You can strengthen your abilities to see details, and to notice conceptual patterns.

Learn to take advantage of what is already around you and to benefit from what others see.

Incorporate this into your own invention process and bring new ideas into the world. **301 Introduction**

Introduction

Seeing is something that most of us take for granted. We open our eyes and the world around us is represented to our consciousness.

Typically, we don't see lines, shading, and forms but objects, people, and activities. Philosophers, scientists, and psychologists have spent millennia reflecting on the process of seeing, trying to describe the mechanisms responsible for this incredible, magical capability.

Few of us take advantage of our seeing abilities beyond using them to move around in the world, to amuse ourselves, or to gather basic information that is important to us. It's not that we don't spend a lot of time looking or appreciating what we see. It's just that we haven't developed the skills to maximize our seeing abilities, or to bring them into a general context for discussion.

In contrast to most of us, some professionals are quite facile at seeing, and are able to make deliberate use of what's around them. Obviously, photographers, artists, bird watchers, and archers rely greatly on their visual abilities and train actively to improve them. Less obviously, research scientists, urban planners, market researchers, and mechanical engineers also rely greatly on their seeing expertise, consciously gathering tools and competencies that extend their visual capabilities.

The intent of this section is to offer some new perspectives on seeing, to provide a basic primer of interesting observations, and to introduce some exercises that can enhance your basic skills.

You Choose What You See

Seeing is something that you choose to do. You open your eyes, you move your head around in a certain direction, you attend to the particular details that you find interesting at that moment. Many of the reactions you have are automatic and not easily available to your conscious control. You turn your head quickly if an object comes at you from the side. Your eyes move along the edges of objects. You notice movement in a scene. Bright lights grab your attention.

These choices are a function of the years you have spent seeing, and your implicit assessment of what is and is not important to you. Because your choices differ from those of other people, you probably notice things that they do not. If you are interested in photography, for example, you will notice the subtle shading of an object or scene. If you are thinking about where to live, you will notice the houses in your environment. If you're afraid of dogs, your eyes will take note of each and every one while you are on a walk. If you are trying to choose a color for a project, you will be aware of all the subtle hues in the objects around you.

You Interpret What You See

Our visual systems respond to brightness, edges, shading, texture, and other sensory and perceptual characteristics of the world around us. Yet generally, most of us do not experience seeing these elements. Instead we "see" a playhouse, or a tall tree, or a cluttered living room, or a pretty girl, or an expensive car, or a talented ballplayer. By the time we are aware of seeing something, our visual processing system has organized it for us and given it meaning – at least enough meaning to identify it and assign it a simple label.

If we look longer and more deeply at an object, we often give it more meaning:

"That playhouse is just like the one I wanted as a child. It looks really cozy inside. I wonder if it is chilly on a rainy day?"

Soon the object or scene we are looking at has become the center of a set of mental associations about our life and our interests. If we look away, we're often unable to remember even the simplest perceptual details of what we've seen, because we have assigned it other, inner forms.

"Was the roof really red, or was it brown?"
"How many windows were in the playhouse?"
"How big was it?"

Part of this phenomenon is a function of the way we process language. We apply words to the things we see, only to find that our perceptions become general

---

66 When I'm looking at the world from a scientific point of view, I'm using a certain mind set that helps me... It's not just looking around and seeing everything... It's taking it in through a structure so I can interpret what I'm seeing in a way that makes sense for my point of view." - Rob Semper

elements implied in the word, while the details disappear. Normally, this use of language is helpful to us; it encodes our worlds in ways that make them easier to communicate about and it provides us with a set of labels that we can easily remember. But when we need to focus on or recall specific perceptual details, we sometimes discover the limitations of this system.

You remember that you saw a golden retriever, but you don't remember exactly how big he was, or the color of his collar. You know there was a bird on the man's shoulder, but you have no idea if it was the right shoulder or the left.

Lawyers have a heyday with this in court. They ask you about an automobile accident, and you calmly describe how two different cars entered an intersection and how the driver of the first car forgot to signal and turned into the other car. You feel as though you are providing invaluable information to establish what happened. Then the lawyer asks you if there were any distinguishing marks on the second car. You say no. He or she then shows you a photograph of the cars which indicates that the right fender was already crumpled before the accident.

You say you didn't notice. How, the lawyer asks, could you be so sure of what happened if you weren't even paying enough attention to notice the fender? You wonder about your own abilities. The judge discounts your testimony, and the jury wonders if you are lying about everything.

These anecdotes describe how your memory records what you see. In actuality, you may have seen everything accurately even to its smallest perceptual detail. But your perceptions were organized and interpreted; that's what you remember now.

Some people do see in great detail, and there are ways to train yourself to become good at this. But typically we use what we know to interpret what we see in any given moment. This is a great advantage, for it helps us navigate quickly and efficiently through the world. By relying on this system of visual labels and stereotypes, however, we can also find ourselves ignoring something of value – or being tricked into seeing something that wasn't there at all.

Visual illusions – typically considered curiosities that tease your eyes and brain – actually provide important information about your interpretive capabilities:

• If you see a line with arrows pointing out at either end, you will judge this line to be shorter than the same line with arrows pointing in.

• When you see a set of circles near a larger circle, you think they are smaller than when you see the same circles near a smaller circle.

• If you see, over time, a sequence of still pictures that are similar, you integrate and interpret the motion of their objects, making them into "a movie."

66 It is with the heart that one sees rightly; what is essential is invisible to the eye..." -Saint-Exupery, *The Little Prince*

Perceptual psychologists have spent decades arguing about the underlaying explanations for each of these illusions, trying to determine just how the human visual system works.

The truth is that you see both what is and what you think is there. Your system is finely tuned to take advantage of past experiences to interpret what you see in context. Usually this helps you out. On occasion it can be a real problem.

To state this in a slightly different way, you don't see like a camera. Your system does not passively register whatever is imposed upon it. That's the good news.

On the other hand, unlike a camera, you can't retrieve accurately everything you're exposed to. That's the bad news. However, you can influence how you see and the kinds of interpretations you make. You can view details as they're needed to make a drawing. You can learn to see perceptual elements carefully. You can also become experienced in viewing the patterns of things, going beyond specific details to acknowledge larger issues and relationships.

The Natural World and Your Constructed World

In any discussion of seeing it is important to distinguish between the "viewer" (the person doing the seeing) and the "view" (the elements that are being seen). Both are obviously important; it's their interaction that we consider when we think about seeing.

What does this imply?

There are a number of philosophical and psychological issues raised by this distinction. The details of these are best found in epistomological and phenomenological texts in philosophy, and in perceptual and cognitive psychological treatises within the field of experimental psychology.

For the moment, realize that what a viewer sees is different for different people and different situations. We are not simple mirrors, but interpreters.

More specifically for our context, notice that what you see is a function of what you choose to see.

- You can choose the views you experience.
- You can look at them with attitudes and intentions you choose.
- You can construct views for particular purposes and see new things.

It is in this basic assertion that one can find opportunities for visual thinking! Let's examine these three activities one at a time.

---

66 When we're designing food packaging, in a sense we're eating with our eyes... We're asking the consumer to taste a product by the way the box looks." -Primo Angeli

Your Choice

Choosing your views can be an intensive exercise itself. For starters, begin to pay more attention. Don't look down at your feet as you walk along; instead take note of all the views that surround you as you move through the world. Most likely, you'll soon find that doing this becomes a pleasurable and aesthetic experience.

You can also be more deliberate, choosing views that are related to a problem you're working on or an issue you're immersed in. If you're trying to choose a dress for a party, wandering through clothing stores is a good idea. A hardware store is a great place to figure out what kind of faucet to put in your new bathroom. A walk in a botanical garden is helpful in planting a spring garden. Certain movies let you work out your own feelings about a relationship gone awry. A hike in the mountains can inspire geological theorizing. These visual connections between the natural world and your inner, constructed world, can lead to practical, real-life decisions.

As you get good at this, you'll find that you can choose less obvious views to represent your situation. Your garden may help you understand the death of your father. Looking at a city's early morning delivery patterns of newspapers, bakery goods, and fresh fruits may inspire your thinking about models of blood flow in the human body. The angles of an ugly building in front of you could motivate a short story about conflict in our modern, urban culture.

With experience, you will make incredible discoveries. You didn't know you were thinking about your mortgage, but suddenly the log floating down the river in front of you prompts solutions about refinancing. Lying beneath a tree, the details of a leaf pattern motivate you to solve a mathematical puzzle.

You can choose to be ready for these kinds of intuitive "leaps," and you can plan ways to maximize their occurrence. You can then use your everyday seeing opportunities to enhance your own productivity and pleasure. And the best part is, it's free and always available! All you have to do is look.

Your Point of View

In addition to choosing what you are going to look at, you can also control how you look. You can determine your attitude and point of view and modify them in order to meet your own goals. As an example, you might choose to see the world around you in terms of the symbolism it conveys, or the metaphors it offers, or the classes of relationships that are evident. Or you can choose to notice colors, forms, textures, degrees of brightnesses, or a range of other perceptual elements.

Similarly, you can choose to focus on content, either in a specific or abstract sense. You can deliberately notice all things related to dogs, for instance. Or you can look for all the different ways things can be opened (with hinges or gears or folds or holes or whatever).

Depending on the filter you select, you can view a playhouse as a symbol of childhood innocence or as a metaphor for passages to adulthood. You can be inspired by the way the sun shines on its surfaces, or the shadows that dance on its painted shutters. You might consider the basic design of the playhouse as a model for affordable housing, or you might look at its porch with an eye for its potential for social activities.

There are many attitudes you can bring to the objects you see in the world. Your challenge is to control these attitudes and use them to take advantage of what surrounds you.

Your Physical Constructions

It is tempting to think of the world as a fixed set of elements that you can walk among, making choices as you go. Yet this is too limiting. You need to realize that you can create elements in your world and then use these to inspire yourself or others.

This is probably most obvious in architectural spaces and other "man-made" environments. Most of us do not spend much time in truly natural environments. Instead we live among the buildings and roads and telephone poles that others have created. And we live in interior spaces over which we ourselves have some degree of control.

On a more abstract level, we live among a range of symbolic representations constructed by ourselves and our colleagues. I "live" in the outline of this *Handbook*.

You "live" in the *Sketchbook* that you are creating as part of your *VizAbility* experience. As you become more and more a part of the visual culture, you'll find that you live in the work you've collaborated on with other people, and in the diagrams and charts you have created to represent your own endeavors.

Even as you are constructing your own physical and abstract visual world, you can continue to use it for new inspirations. You can use drawings and diagrams to see new things and to move your thinking along. By making a chart of a new project, you may discover that you will not meet your deadline or that you need certain unanticipated resources. By sketching a new piece of scientific equipment, you can become convinced that it won't work the way you intended. In each case you are looking at familiar representations on arriving at new understanding.

Your task is to get good at this – to learn what kinds of constructions in your world help you see what you are doing. As we've often noted in this discussion, you need not be a passive viewer; learn to see actively in order to benefit from what's around you, in both natural and constructed environments.

Perceptual Elements of Visual Form

Once you take advantage of seeing actively, you'll realize that all sorts of visual nuances exist. These are well-known to the artist and photographer, but they are somewhat obscure to most of us "normal" viewers.

---

66 The importance of doing science is actually seeing, and there is a difference between looking and seeing. When you look at something, you see what it's like, but when you see something, you actually see what it's made of...what's behind it."
-Rob Semper 301 Overviews

For instance, most of us are oblivious to the subtleties of color. We notice them only on random occasions, such as when we try to match a paint color for a wall, or find a shirt and a pair of trousers that go well together. Texture is an important element of objects, yet we usually ignore this attribute until we have to draw something. We rarely notice shadows and shading, as is obvious when we get our photos back from the developer! How many of us have taken pictures of animals at the zoo, only to find most of them obscured by bars we didn't even notice when pointing the camera?

The list of perceptual attributes includes:

- color
- orientation
- shape
- brightness
- superimposition
- similarities of form
- symmetry
- negative space
- size
- spatial relationships
- texture
- connectivity

All of these things we can choose to see or ignore. You can take art classes to familiarize yourself with these and other visual attributes of the world, to train your eye and to assist you in drawing and other constructive activities.

Art historians study how different artists use these elements to show what they see. Computer vision experts, in designing robots who move about by processing visual data, also pay attention to different features of the visual world, to gather clues about what objects are in the robots' paths. Perceptual psychologists study how the human visual system highlights the features that are important in different situations.

Begin by Noticing

Look around yourself right now. For a minute or two, just notice the brightness in your environment. Pay attention to where the light sources are.

Look at an object in your environment. Is there a direct source of light on it? Or is it mostly reflected light?

How do different surfaces in your environment reflect the light differently? Do you notice any interesting shadows? Take your time.

Now think about color, and pay attention to the colors around you for a few minutes. What is the most prominent color in your vicinity? How many basic varieties of this color do you see? Are there any blues nearby? What about dark browns? Are there many kinds of "off-whites"? Look for colors on paper; how do these compare to the natural objects around you?

Try this for each of the perceptual attributes listed. Later, as you wander casually through the world, try these exercises occasionally. Integrate these "seeing sensitivity" moments into your daily life and discover what you have been passing by all these years!

66 Sometimes, if you're stuck or working on something, you have to wander around and look at things, see what they might remind you of..." -Denny Boyle

Transformations in Two, Three, and Four Dimensions

In addition to simply observing your world, learn how to make deliberate changes in what you see. Move toward objects and notice how their apparent size changes. Note how the amount of detail increases. Watch how the relationships between objects shift as their superimposition changes.

Although the objects remain constant (e.g., the refrigerator doesn't turn into an elephant or the coffee mug into a chair), the appearance of each object varies greatly depending on what you do.

There are actions you can deliberately perform that transform how you see an object; you can change the lighting, for example, or you can look at the object from a different angle. Gradual changes in the environment, like the sun coming up, also affect what you see. Your task is to acknowledge these changes and to play with them. Entertain yourself in your next boring meeting by imagining just how everything might look upside down. Look at the shadows on a friend's face instead of the features; now change your view and notice the angle of his nose and the color of her eyebrows.

Here are some transformations that you can explore to enhance your seeing capabilities:

In two dimensions:
- rotate, locate, identify, scale, combine, notice, substitute, omit

In three dimensions:
- fold, change light source, change point of view

In the time domain:
- bounce, stretch, expand, compress

You should both look for these kinds of visual changes in your world, and do things yourself to make them happen. Enjoy silly wind-up toys; crank them up then watch carefully as they move. Throw some things and watch as their appearances change. Dangle objects on strings in front of lights and marvel at the shifting shapes and shading and sizes. Try to take special notice of all of the details that, as an adult, your mind has chosen to ignore, and let yourself feel the pleasure of incorporating them in your everyday consciousness. Become a child again and actively explore your visual world!

---

66 I oftentimes like to take double-exposure photography, because I'm always looking for a different relationship than I can get naturally...that's the way I try to see, trying to create new relationships with existing things." -Jeff Zwerner 301 Overviews

Visionary: One who is able to see trends, patterns...to see the whole in order to guide the details of everyday life...

Seeing and Drawing

Drawing is an important activity that can help you learn to see again. Begin to draw things to focus your vision and test what you are seeing. Glance at an object or scene around you, then draw as much as you can remember in your *Sketchbook*. Glance up again and gather more details, then continue your sketch. Continue the cycle of seeing and sketching until you feel your sketch is complete.

As some of you try this, you may become frustrated at "not knowing how to draw." You might wonder: "How can drawing help me learn to see if I haven't yet learned to draw?" If this does apply to you, remember – it doesn't matter what your results look like. You are not creating a work of art, something to hang on the wall and show off (or even to stash in a drawer). You are drawing in order to see. What matters is that your attempts to draw are directing your eyes, and capturing visual information.

Of course, as you use drawing to see more actively, and as you become more observant, you will most likely produce better drawings. Don't be surprised when this happens. For as you become sensitive to visual elements, you'll increase the mental resources you need for drawing; for instance, you'll develop a better sense of shape as you draw an object or an awareness of colors as you capture a scene.

Seeing and drawing are amazingly intertwined. Each can help you bootstrap skills in the other arena, sometimes the two may even become indistinguishable. You will see many levels in the things you choose to draw, and you will only draw well what you can really see. Begin the cycle now with an emphasis on Seeing.

Seeing and Imagining: From "What Is" to "What Might Be"

The dimensional transformations described in the earlier section assumed a constancy of objects and scenes. We subjected an object to different light sources, or different angles of vision, but the object itself remained the same. What happens if we relax this constraint and deliberately change the basic elements in our perceived worlds? What if the refrigerator turned into an elephant, or the coffee mug into a chair?

More practically, what would the playhouse look like painted red instead of white? How would milk pour from a carton with a metal spout? How would a cake stay fresh in a shrinkwrapped package? How might the kids play on a new slide? What would life be like if you went to law school? Who would use a community center if it were built?

What happens when you consider "what might be" as well as "what is" when viewing the world?

When you take this approach, you have truly entered the domain of "design," a domain that is typically identified with a particular segment of society (e.g.,

Practice your transformational abilities with Transformations 621, Block Builder 623, and WarmUps 780 on the *CD-ROM*.

"designers") but that actually encompasses a much broader context. You have combined your ability to see ("what is") with your imagination ("what might be") and opened opportunities to create new forms and approaches in both your inner and outer worlds.

Once one takes this approach, one is involved in "inventiveness" if one is in engineering, "creativity" in the arts, "problem solving," "problem finding" and planning in everyday life, "strategic planning" in business, "research" in the sciences, and "hopefulness" in the arena of social issues.

The perspectives involved can be concrete and perceptual, such as seeing the playhouse and imagining a new coat of paint. Or they can be abstract and perceptual, such as considering which elements you need to make a "comfortable" house or an effective school.

Learn to notice these opportunities for change in the world around you. See "what is" with such intensity that you can imagine ways in which it could grow or change or evolve.

If actually accomplishing all the changes you're capable of imagining seems overwhelming, set these issues aside as second-level considerations while you're tuning up your visual system. You need to begin the seeing process with a sense of freedom, allowing your visions to develop. Later, if it is appropriate, you can apply practicality and ingenuity to carrying out your ideas in concrete form.

Observational Competencies

As you interact with your visual world, you may be surprised at how much fun you are having! With a simple turn of the head, or a glance over your shoulder, you may glimpse something that is startlingly beautiful. With a certain viewpoint or particular intention, you can discover productive new approaches to your work through your everyday world.

Knowing how to see offers some incredibly practical advantages. It can help you remember what is going on around you. It can help you stay out of trouble (e.g., avoid the falling rock or the runaway truck). It can help you to identify crooks or decorate your house!

It can also help you get jobs that require special observational competencies. Bird watching, sports broadcasting, experimental botany, marine biology, cinematography, medicine, umpiring, house painting, air traffic control, and fire watching are just some of the careers that come to mind. For although most of us never get formal training in seeing, observational competencies are central to a large number of professions.

Seeing well can also motivate a range of problem defining and problem-solving activities in the more abstract fields. Mathematicians can observe patterns in data, product designers can see opportunities in old products, and market researchers can predict the success or failure of products before they are even built.

66 It's very important, as a designer, to have the ability to see problems. To see what irritates not only you but other people. To see what other people struggle with but don't realize is a problem. It's sort of this ability to recognize needs."
- Denny Boyle

Enhancing Your Seeing

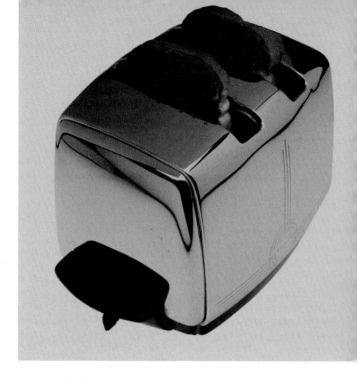

You can improve your seeing by noticing what is around you, by completing some seeing exercises and by talking with your friends about how they see.

Amongst the many possible ways to enhance your seeing are the following, each of which corresponds to an activity on the *CD-ROM*.

These abilities span a range of levels, from sensory and perceptual skills that typically feel automatic, to more deliberate cognitive activities that you can control.

Learning to see quickly helps you become skilled in the acquisition of information. Seeing over time helps you recognize the dynamic properties of images. Noticing details adds a range of dimensions to objects, both at perceptual and conceptual levels. Increasing your ability to see flexibly at higher cognitive levels can make you facile at all kinds of tasks. Learning to gather meaning from what you see expands your visual perspectives, multiplying potential approaches to tasks.

You can find ways in which to combine these "levels of seeing." Sensory details can affect cognitive strategies and vice versa. Let's take these approaches one at a time by looking at the activities on the *CD-ROM*.

Seeing Quickly

The ability to gather information quickly from a scene is extremely well-developed in the human visual system and has probably been studied more than any other visual attribute. Perceptual psychologists are forever flashing quick views to experimental subjects, who respond with a range of judgements from "Yes,

I see it now" to "Yes, I would love to buy this." Traditionally, this was accomplished with a machine called a *tachistoscope* that controlled viewing time, intensity and other variables. Today, computers control these presentations.

Students in studio art classes are frequently shown brief views of objects, which they are then asked to draw. In some cases, this is accomplished with a slide projector or by holding a card or other obstruction in front of an object. In other cases, students are simply asked to look up for a certain time and then look down and draw.

The *CD-ROM* duplicates this exercise in its FLASH SKETCHING activity by providing you with brief flashes of different images. At the simplest level, your task is to practice seeing the images, learning to see as much as you can in the fewest number of flashes. A range of different images is included for you, from basic silhouettes to simple objects to more complex human forms.

FLASH SKETCHING is an exercise that you can do again and again, even with the same images. It is also an exercise you should take some time to experience. You might look at the same object over and over again, for example, assuming a different angle of looking at it each time.

It is also an exercise that you can modify to train different aspects of your visual skills. For instance, you might change what you "see" as you look at an object. First, just try for recognition. ("It's a telephone!") Then try describing the object in detail. ("It's a black phone, with a shiny surface and a round dial with fingerholes next to the numbers.") Next assign the object some level of meaning. ("I don't really like black phones, they remind me of the doctor's office when I was a kid.") In the first two descriptions you are focusing primarily on perceptual details, while in the third you are concentrating on emotional and symbolic elements.

Extend this experience by drawing the images after you glance at them. Flash an image and begin to draw it. Then flash it again to elaborate your drawing. Keep flashing the image until you are satisfied with your drawing.

Flash Sketching shows how a flower might be drawn and in what sequence. Notice how you complete your drawing. Overview then details? Top then bottom? Try a number of strategies and notice how they affect both your seeing and drawing. 311

Choose an object in your environment. Glance at it briefly and then try to draw it. Continue to glance, then draw, until you've completed your drawing. Put together a stack of magazine photos or shapes, and repeat the exercise with these.

Reflections

As you explore this exercise, take some time to reflect on your performance as well as your experience. For example, how does the number of flashes necessary for you to "see" an object vary? According to your experience (first image vs. twenty-first image)? According to the object (simple shape vs. complex shape)?

What happens when images are recognizable? Does it help you draw them? Or does it hinder your perception of detail?

What about personal meaning? Is it easier for you to visualize/draw dancers because you have danced yourself? Or harder because you have never danced?

What about other tasks: Is describing the object easier or harder than drawing it? How does simply recognizing the image compare with either drawing or describing it? Consider the different ways you look at images for different tasks. (For example, you will probably look at more perceptual details when you choose to draw an object.)

As you experiment with FLASH SKETCHING, you'll become aware of many basic processes of the human visual system. You will probably notice, for example, that your eye attends well to the edges of objects, particularly in brief flashes. You'll find that basic forms are evident quite quickly, whereas details require multiple flashes (unless they are particularly

prominent or unusual). You may find that your knowledge about something affects your ability to see it clearly. And you'll probably find that your past experiences affect what you see (you'll have a much easier time "seeing" elements with which you are familiar).

You may find that you quickly forget the details of images that are familiar to you. You might notice that a phone has a handset but the details of the handset are not obvious to you. Or they may seem to resemble your own phone and not the one shown. You may find that your drawing is somewhat generic, resembling a stereotypical image of a phone rather than the specific model you were shown.

You can train yourself to see better in ways you consider important. For as you become conscious of how you see, you can extend your new abilities to your daily life. For instance, apply what you learn in FLASH SKETCHING by looking up quickly at a meeting and then trying to sketch what you see, or trying to catch details as you ride by in your car.

Seeing Over Time

A camera is designed so that available light reflects through a lens system and then leaves a trace on the film inside. Early theories about the human eye suggested a similar process – light imposes an image upon the retina and enables us to see. Investigations in the last century, however, have shown us that this passive receptor theory is far from the truth.

Instead, it turns out that our visual systems are active participants in the seeing process. Our eyes move rapidly, gathering bits and pieces from the world to be "assembled" later. These eye movements are not random; they are set up to select and prioritize information. For example, our eyes naturally exhibit enhanced responses to edges. The non-foveal region of the retina is extremely sensitive to motion, providing intense response to sudden change.

Our visual systems are also keyed to observe changes across a broad temporal plane, in order to form predictions. We watch storms build over time, for example, to get a sense of the weather patterns we should expect. We watch a fly ball sail into the outfield in order to consider how to catch it, or to place our bets on whether the fielder will get it.

Sometimes the course of these changes is fast, other times it is slow. Often we try to vary the durations of these events in order to see patterns that are not readily available to the naked eye. For instance, we might film high and low tide levels on a beach over the period of a month – then speed up the film to watch the water's response to a lunar cycle in just a few moments. Or we might take a movie of a swimmer doing the butterfly and then slow it down to better observe the patterns of the stroke.

Have a friend trace letters, shapes, or symbols on your hand. Your task is to identify these forms.

Look at the world around you. Pay particular attention to the things that move.

Try the EYE TRACKING exercise on the *CD-ROM* to extend your experience with this basic set of phenomena. Your task is to recognize a letter, shape, or symbol as it is traced on the screen over time.

Some of you may have memories of lying on the beach as a child and playing a guessing game as friends wrote words on your back. Or you might recall "waving" letters in the air with lighted sparklers on the 4th of July and having your friends decipher what you wrote.

Although the basic sensory mechanisms in the above examples differ from those in EYE TRACKING, (e.g., your back is sensing the movement through touch, and a bright sparkler causes more persistence of vision than a dot on a computer screen) the basic experience of consciously "seeing over time" is common to all these situations.

As you master the letter forms in the first exercise, try the shapes and then the symbols. Here the number of possible alternatives increases because of the complexity of the images. Try to keep looking at the dot, deciding what the pattern is, and then – and only then – check the alternatives to confirm your answer. Or, sketch your answer first, then select it and see how you did.

In this exercise, the sequence of viewing is controlled by the computer. But what happens if you control the sequence? Try this: Make a small peephole in a piece of paper, or roll the paper into a "telescope." Then look at the world by moving your viewer over scenes and objects in front of you. Notice how long it takes you to figure out what you are looking at. Alternately, have a friend select an object and put it on a surface in front of you. "Pan" over the object with your viewer until you can guess what the object is. Now try panning over a scene on a piece of flat paper, such as an illustration from a magazine.

You'll find that, with experience, you get better at choosing views that permit easier identification; you'll also come to appreciate how your visual system unconsciously assembles patterns from large sets of small elements (an activity it does all day long).

As you get used to this, switch your technique. Instead of moving your viewer, move the scenes or objects. Place a photograph or illustration in front of your peephole or telescope and hold the viewer still, moving the scene in front of it. Or have a friend choose an object and move it in front of you. How long does it take you to recognize something this way? Try this for a number of different scenes and objects.

How do these two methods (one in which the viewer moves and the other in which the viewed moves) compare? Which seems most natural? Can you learn to do both? Consider what this reveals about your visual system.

Reflections

Which of the EYE TRACKING exercises is the hardest for you – letters, shapes, or objects? Why do you think this is? Is it the complexity of the forms or the range of elements that you must distinguish between? How did your sense of anticipation help or hinder your attempts to identify elements? Did you find that you knew what something was well before it was completed? Or did you need to view the entire form before you had any clue to its identity?

One variable that we could have built into the EYE TRACKING exercise (which we discussed but decided not to implement) would have been differing speeds for the moving dot. We manipulated this variable quite a bit in preparing this exercise, trying to find a speed that on most processors would make the exercise difficult enough to offer a degree of challenge, yet not so hard that users would become discouraged. The results of our experiments were interesting. In most situations, moving something faster makes it harder; in this exercise, moving the dot faster made it easier to notice the patterns and to see the display as one integrated form. Would you have predicted this? What does this say about how your visual system works?

Another thing we experimented with in preparing this exercise was the size of the dot. Here again, we tried to find a size that would make the exercise compelling without being boring or frustrating. Ultimately we even considered giving the user control over this variable but we decided against this for practical reasons. What do you think your experience would be if the dot were larger than it is? Smaller?

We found that when we made the dot larger we began to see the lines as solid; the dots integrated over time and no longer appeared to be individual units but instead connected to form lines. What does this make you think about your visual system? And how might you take advantage of this in your work?

Most activities of seeing over time are quite unconscious for adults. Our visual systems respond quickly and automatically to dynamic displays, readily generating interpretations. Spend some time practicing this, however, and you'll discover that you can improve your efficiency. In addition, by thinking about this set of phenomena, you may find a range of useful techniques for presenting ideas that make use of time-related aspects of seeing.

Noticing Details

We can learn to focus our attention on seeing general patterns over space or time. Alternately, we can focus on very precise visual details. In HIDDEN PICTURES you can gain experience in viewing the details of images. Your task here is to identify where a small, square image fits within a larger picture.

See how quickly you can find where the small square belongs. Try this exercise with a friend and see who is quicker. Talk to each other about your strategies. Improve both your speed and technique with practice.

Finding details within a scene is a common task. All kinds of games have evolved around it, including the recent rash of books and games in different flavors. It seems to be a compelling activity, and one that has great inherent reward for "the find." You search and search, then – Eureka! You find it! You experience a sense of great satisfaction, and the tension is finally dissolved.

Occasionally, you may notice as you try the HIDDEN PICTURES exercise that you locate the small square almost automatically, before you're really aware that you're looking for it. As you practice, you'll probably find that this occurs more frequently. In those instances when you simply can't find the small square, click on "Grid" to place an overlay on the large picture. This simplifies your task a bit as it breaks the overall image into smaller segments. It's easier to scan the contents of each small square and quickly eliminate certain regions of the picture.

331 Hidden Pictures Introduction
332 Look for Details in Black and White Pictures
333 Try Some Color Pictures

Make two copies of a picture from a magazine or some other flat art. Cut one copy into small squares. Select a small square and try to find it in the other, intact, or picture.

Sometimes you'll know that you have found the small square, but you will have little consciousness of what is depicted in either the overall picture or in the small square. This is because you've focused so well on the details that you haven't integrated the elements of the scene into a meaningful interpretation.

On other occasions, you will be very aware of the nature of the scene – red and green peppers, white clouds in a blue sky – and you will use this information to help locate the small square.

Depending on the situation, being able to call on each of these strategies (either focusing on the parts or on the whole) is important. You might try practicing both of them, so that you can have each available when it seems appropriate.

Generally, in this game, you need to ignore the contents of the picture in order to be effective. You need to be aware of perceptual details, not cognitive ones. Even the most luscious looking strawberry in the world should be viewed as a flattened combination of color, shading, form, and texture.

In HIDDEN PICTURES it's more efficient to focus on perceptual detail, not only because of the nature of the task, but also because of the type of pictures used. We deliberately chose images whose parts are similar, ones that can be distinguished only by spatial relations and visual elements.

You can convert HIDDEN PICTURES into a patterning exercise, if you wish. Instead of first finding the small square, try "naming" the small square first and then looking for it.

• A round brown object with a dot in the middle.
• A cloud bottom that curls to the left.
• The left hand side of a twenty-five cent piece.

In HIDDEN PICTURES we begin with an arbitrary detail that has been plucked from the main image; it typically has no coherent identity or behavior that helps us in finding it. In naming it, you give it some meaning that may help you find it.

Try a number of naming tasks, learning which kinds of names help and which don't. "Feel" the difference between matching perceptual vs. cognitive patterns and descriptions. This will help you to develop your own visual viewing strategies.

Reflections

Which pictures were easiest for you? Which were the more accessible, black and white or color? Did you become quicker with practice or did you get worse, with more examples wearing you out or making you self-conscious?

Try the exercise a number of times, keeping track of your progress. You might want to time yourself on the task, just to see how fast you are. You might also play against another person to help to focus on your speed.

You should also try to transfer this kind of detailed seeing into your daily activities. Notice perceptual details in the world around you. Jot down notes about them. Learn to identify and remember compelling forms. Do quick sketches of details that have never concerned you before – the detailing on the windows of a large building near you; the shapes of the leaves on the trees; the elements of a friend's dress on a particular day; the molding on a doorway; the shadows that you see on the ground. Tune your eyes to notice details for awhile and move out of the realm of everyday seeing.

You can also use HIDDEN PICTURES to enhance your drawing ability directly. Simply turn on the "Grid" and copy its pattern into your *Sketchbook*. Then try drawing a picture by sketching in one square at a time. This technique, used frequently in drawing workshops, can help you learn to render details, details that you

normally aggregate without thinking about it. It is an effective technique for breaking down a formidable scene into manageable units.

Seeing Flexibly

In today's primary schools, most students come into contact with "attribute blocks." These blocks give young children direct practice in sorting objects that are like and unlike one another, a skill that has been associated in recent research with higher learning or "critical thinking" skills.

As adults, however, most of us do not have experience with these kinds of blocks and may even think of them as "child's play." Yet sorting tasks are central to many intelligence tests and are good indicators of general abilities in concept development. In fact, artificial intelligence researchers have experimented extensively with designs for computer programs that can sort objects according to their attributes. They have even developed more advanced programs that can guess the categories into which objects are sorted by looking at their attributes. One of the most important things we can learn from this kind of research is that sorting and categorizing abilities are non-trivial; indeed, they are central to the most basic reasoning tasks.

A highly critical element of sorting is the ability to group two objects together to fit one category and then to separate them to fit a different category. For example, Mary and John are both highly competent physicists and are therefore both members of the category "competent physicists." But Mary is also a member of the "female" class or set whereas John is a member of the "male" category or set.

At some level this is obvious, and you may wonder why so many noted philosophers and social scientists have given their attention to basic set membership. As one explores this, however, the relationships get more complicated; and some interesting generalizations and inferences merge. Some of these are made with correct logical inferences, others are common but quite illogical.

For example, many people will state that there aren't any competent female physicists, based on the notion that most of the physicists they meet are male. If most are male, then what are the chances that the really competent ones are women? Even if they meet Mary and acknowledge that she does happen to be a competent, female physicist, they will continue to generalize that competent physicists are male. Although their estimations will be correct most of the time, it is clearly an overgeneralization and one with potentially harmful and inappropriate ramifications. It might mean, for instance, that young women won't go into physics, or that male physicists don't think of women in their field when they are determining awards.

Beyond the substantial and damaging social consequences of such overgeneralizations, this type of incorrect reasoning can prevent us from seeing opportunities. New theories are often overlooked because people have neglected to notice certain differences among elements in the same category.

Have a friend secretly group a set of objects or pictures into two categories, making sure some objects do not fit into either category. Now look at each object and guess which category it is in. Your friend can give you feedback as you discover the categories.

Likewise, new product possibilities can be missed when attributes are clustered or identified too rigidly. In designing the Duo Dock for Apple Computer Inc., for example, Denny Boyle and his colleagues questioned general assumptions about computers. They looked at the differences between portable computers (which have batteries, can be carried, and are lightweight) and desktop computers (which have large screens, are networked into external services, and attach easily to multiple devices). By questioning this division they created a new class of machine that is both a portable and a desktop, since the portable unit "docks" easily into a desktop configuration.

The ATTRIBUTES exercise provides you with an opportunity to notice and develop your own grouping abilities. The computer has been programmed with sets of letters, shapes, or objects. These sets are not apparent to users; all you are given is one item at a time. You are asked to group them into three different groups: "Set I," "Set II," or "Discards," which belong in neither "I" nor "II." As you do this, you are deducing the correct categories.

Try this exercise and notice which categories seem straightforward and which are difficult or confusing.

Reflections

Indeed, when the first element comes up – a letter *K* for example – there is no reason for you to know which groups it fits. You have to choose a category without any information, gathering data for yourself as the system accepts or rejects your guesses.

This action is pretty daring, not unlike taking that final stab at solving an important problem. Your reflection in this particular ARC cycle is pretty straightforward, however, since the computer lets you know if you are right or wrong. If you are wrong, then you have two other choices to make. Even if you are correct, realize that the element may actually fit in both groups, not just the one you selected. This does not pertain to items for which you selected the discard pile correctly.

Your task is to continue acting and reflecting to formulate your hypothesis about the nature of the grouping, learning from your previous choices.

Some of the examples in ATTRIBUTES will prove easy and sensible. Hopefully a few will be a little mindbending, forcing you to try a number of approaches in order to guess the categories correctly. For "trying things out" when they don't fall into place, and "looking at it a bit differently" in order to solve a problem, is the entire purpose of this exercise.

Most of the categories in ATTRIBUTES are fairly conventional. We discussed and experimented with odd groupings – "edible" vs. "contains wood," for example – but found many people confused or uncomfortable with categories that were less than diametrically opposed. So we settled for contrasts, with a random element to keep you on your toes.

Consequently, it's your job to extend your sorting abilities into less obvious categories. Try it on-screen, ignoring whether the computer accepts your answers or not. How many different categories can you come up with for each set of letters, shapes, and objects?

Try using some very subjective judgements – like vs. dislike, comfortable vs. uncomfortable, fun vs. irritating – or some very unrelated attribute sets – pleasant vs. red, fancy vs. metallic. Extend your general flexibility of vision beyond the examples provided.

Then apply this new skill to random objects in the world. Does your fork at the breakfast table go with your computer on your desk, for example, or are they in different categories? If they are in the same category, just what could this category be? Keep working on this skill. It can help you unblock stereotypes in your seeing and your problem-solving. You can learn this skill to frame problems in new ways that might even be silly – but that might lead you to solutions of engaging problems.

Seeing Meaning

How do people who are experienced visualizers describe the way they see? Do they pay particular attention to details? Are they always looking for engaging forms in their environments? Are they extremely conscious of the seeing process, focusing on it all the time? Do they have any sense of how they developed their capabilities?

We talked to a number of proficient visualizers – cinematographers, graphic designers, market researchers – to gain insight into these questions. We found that very few of them were articulate about their visual processes. Seeing is just something they do, and they weren't particularly interested in trying to figure out how, or explain it.

So we tried another tack. We asked these people to look at some specific objects and to tell us what they saw. This resulted in an avalanche of perceptions, a rich diversity of visual experiences and approaches.

In SEEING OBJECTS you can explore the same objects on the *CD-ROM* we showed our experts. You can click to rotate an object and see its details. It's not quite the same as having the object in your own hand, but it does give you a general sense of shape, color, material, and so on.

Take your time in looking at these everyday objects – a teapot, a milk carton and a pen – before you read about or listen to what others see in these forms. Although you are probably familiar with objects like these, and may find few surprises in them, take the time anyway to write your descriptions in your *Sketchbook* for future reference.

Our "expert seers" seemed to love looking at these forms, touching their surfaces, and pointing to certain features as they described each object. We learned a number of things from these people, and we've summarized the major points in the paragraphs that follow.

People Construct Meaning from What They See

People bring contexts to their seeing, and they derive meaning from the simplest objects according to this context.

For example, Denny, a product designer, gazed at a milk carton and saw a successful piece of packaging. Erika, a nine-year-old, saw a bird feeder. Alison, a teacher, saw an opportunity to teach arithmetic to her second-graders. Rob, a physicist, saw a very regular form with particular properties.

Each person's background greatly affects how he or she considers even the most everyday objects. Being involved in a particular profession, each has learned to see the world through special visual "filters." This can be valuable, extending one's capabilities in one's craft. Conversely, these filters can also be limiting in certain situations, making one less effective.

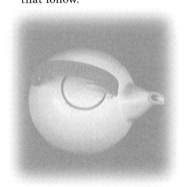  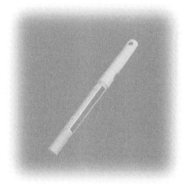

Choose a few different objects in your environment. Ask a number of friends to write down descriptions of these, and write a description yourself. Compare the different descriptions.

People See Differently

People view the same object in extremely different ways. Listening to various descriptions, you would hardly believe the same object is being described.

No wonder we all get into arguments about the most basic things! If our definitions of even simple, everyday objects contain so much disparity, it's not surprising that political, social, and economic discussions become so heated. The miracle is that we ever see anything the same way.

Of course, the flip side of this is that you can benefit greatly from these varying perceptions. The more perspectives you have access to, the more flexible your seeing capabilities can be. Learn by listening to these different views. Among other things, it is a foundational element for team building; savvy team leaders often surround themselves with people who have different perspectives so that they can benefit from a range of ideas.

As a producer and a designer, one can become very narrow in the consideration of a product. However, if one expects others to pay for a product, enjoy it, or benefit from it, then it is crucial to gather information about other people's views. Marketing researchers work toward this end by talking to a range of target audiences about the product in order to anticipate its acceptance or rejection.

People See Different Details

Our visualizers also noticed very different perceptual details in the objects. Denny focused on the handle of the teapot. Based on his experiences with heat transfer, he wondered why particular metals chosen for the handle were used since they absorbed rather than rejected heat, and thereby exposed consumers to potential discomfort.

Jeff, a graphic designer, focused on the fin on the top of the teapot lid and the "fishy" elephant spout. As a market researcher, Brad commented that this was a "nice authentic teapot," that it had a "nice handle that you could use to serve it with," and that it was "attractive" – all attributes that a marketing person would see as potential for advertising and consumer interest.

Each person seemed to hone in on one or two features. This makes us wonder just which aspects were really in the teapot, and which emanated from the viewers' personal biases. It causes us to acknowledge the "constructed" nature of even the most basic perceptual features, and helps us understand the importance of analyzing information in ways that emphasize the viewer as well as the viewed.

66 I think that everybody comes to seeing with a different outlook, and sometimes you're not always on the same level or in synch. So I think that has something to do with the way you see; it's not just with your eyes, but your mind, your body, and your soul." -Nancy Zeches 301 Overviews

People See Abstractions

Consider the depth of the metaphors and abstractions our viewers saw in simple everyday objects. Nancy saw a warm group of people and a comfortable setting as she looked at the teapot; she saw all the possibilities of human communication in a simple pen. Frank saw the teapot as the artifact of a dominant white culture, one that felt irrelevant to his experience as a Black American. Rob and Jeff revealed their likes and dislikes regarding pens, even though they were not asked for these kinds of evaluative comments.

The ability to make abstractions has great advantages, providing the basis for most of the art and poetry in our culture. But it also can get in the way of concrete discussions. You might think you are looking at and talking about a tree, only to discover that the debate is really about the future of ecological movements, or endangered species.

Reflections

Now that you have listened to other people describe these objects, consider how your own perceptions compare to theirs. What kinds of things do you notice in everyday objects? At what levels of meaning do you describe things?

You might try doing a number of different descriptions for each object: long and short; perceptual and meaningful; pragmatic and romantic; and so on.

Also, try viewing the objects as someone else: as a product designer; as a poet; as a teacher of young children; as an accountant.

Learn to change your perception of things, and to acknowledge the range of perceptions that others bring to objects. Then spend some time thinking about the implications of this in your everyday life. Consider how important the initial framing of a work project is to its success, or how crucial your view of a problem is to its solution. Think about how to acknowledge (and sometimes dismiss) the differing perspectives others have on simple situations. Reflect on how to take advantage of your visual flexibility and your ability to assign meaning to simple things, without getting stuck in your own point of view.

Moving On

In this section you have learned a number of basic distinctions about the seeing process and have become aware of the complexities of this visual skill. You have practiced your own seeing abilities, and you have developed some basic perceptual as well as reflective skills. You have come to view the everyday world around you a bit differently, as a source of inspiration and enjoyment, and as an ever-changing resource.

Perhaps you look around you now, instead of looking down at your feet! Remember, the world is brimming with opportunities for playing with your own seeing processes. Take every opportunity to develop them, by yourself and with others.

Like the requirement to Act in the ARC cycle, it is important to take what you see into the world for conversation and analysis. It's time to think about drawing. It's time to learn how to represent what you see in a form that makes your visions clear to others and explicit to yourself. For as you are able to draw what you see around you, you are able to enhance your communications abilities and your abilities to analyze what you notice. You also will become attuned to a level of perceptual detail – detail you will need for your drawings – of which you may be currently quite oblivious. These details will become very important to you in giving life to your drawings and in considering new ideas that you develop.

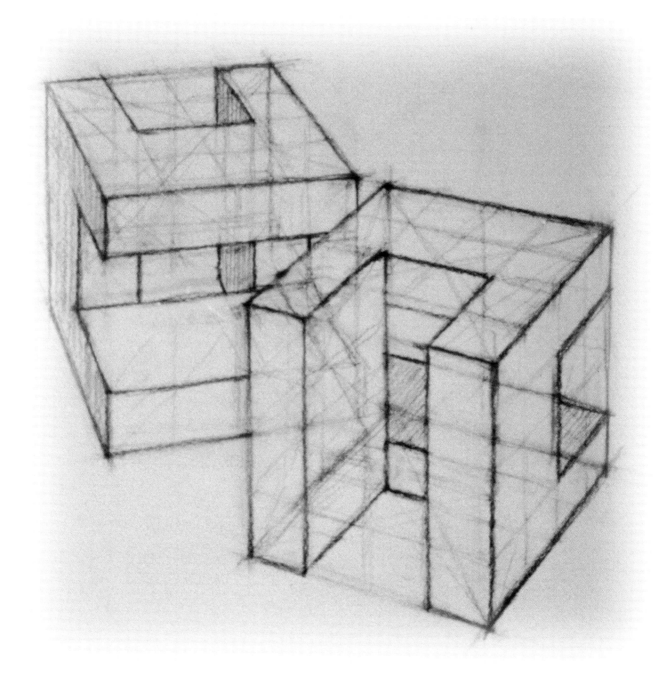

Drawing

Learn to think of drawing not only as an artistic ability, but as a primary communication skill

that enhances your invention and stimulates productive conversations.

It is the primary tool for making ideas visual. It invites discussion and refinement.

Gain confidence in your drawing abilities by learning some basic techniques...

appreciate nuances in lighting and shading, which can bring objects to life.

Gain a fundamental understanding of perspective, which allows you to show objects in space. **401 Introduction**

Introduction

Simple drawings in their many forms – pastels, charcoal, watercolors, pen, pencil – are a wonderful part of our life and culture. They are pleasing or thought-provoking visual representations of the world.

We hang them on our walls, give them as gifts, and purchase them as investments. We are drawn to those that suit our personal tastes, and our interpretations of the world around us or within us.

Many of us are collectors of drawings. But how many of us create them?

Drawing is viewed as a mysterious ability in our culture, one that is reserved for select, talented individuals, the "high priesthood" of art, as it were. And indeed, many individuals seem to have an innate knack of expressing their ideas and visions on canvas or paper, and are appropriately attracted and involved in formal artistic disciplines. Consequently, for most of us, drawing is relegated either to our early school years or the hobbies of late adulthood, as if it were relevant only to the beginning and end of our lives. It is a skill that is approached lightly or not at all during the bulk of our education or professional activities.

But excluding people from the experience of drawing because they are not artistically "gifted" is like excluding people from speaking because they are not great orators or from writing because they are not first-class novelists. Drawing is not just a way to produce art, reserved for those talented in techniques and materials. It is a critical skill for bringing ideas into the world, and a tool for better learning and communication. It is also a process that you can engage to center your attention or stimulate your imagination.

❝ Drawing has a history in our culture of being associated, particularly in schools, with realistic rendering. So children get very inhibited in second or third grade when their horses don't look like horses, or their sky is drawn an inch wide across the top instead of all the way down to the horizon." -David Sibbet

Reason for Drawing

Each of us needs a way to bring physical objects "to the table" for conversation, whether we're discussing a car design, a detail in a house, a new video game, a teaching technique, or a beautiful waterfall. Let's look at several important reasons you might engage in drawing.

You can make use of drawing in order to further your understanding. When you're trying to work out the details of something in the world, (fixing the locking mechanism on a door, or considering the damage on your new car) a drawing can help you reflect upon the issues. It can be useful both when you are in the presence of an object and when you are away from it. The concrete quality of a simple rendering can help you review a situation and determine how you feel about it.

Drawings are also useful when you're trying to imagine something in the future – when planning your spring garden, for instance, or designing a piece of experimental apparatus. First, you can check and double check your drawings to note deficiencies in your plans. Later, you can use your drawings to implement these plans, calculating the amount of seed or number of screws, or the amount of labor required. The process of drawing lets you link your inner and outer visions, seeing "what is" and drawing "what might be."

Another reason to draw is to share your ideas with friends and colleagues. Your drawing then becomes the focal point for a series of communications, articulating what you know and don't know about a situation.

As you examine a drawing with someone else, you can point to particular elements to explain what you mean, and you can provide more elaboration on issues that are challenged. You can elicit feedback and make changes. As this process evolves, your drawing comes to reflect the interaction between members of the group; ultimately it represents the current best approach to a problem. Hence, your initial drawing is the "opening statement," whereas the final drawing is a mutual collaborative solution.

Collaborative drawings have a neutral quality that can be very effective. By focusing on a drawing, people tend to concentrate on the ideas on the table, rather than the different personalities and social dynamics involved. They also keep discussion focused on specifics rather than on vague and nebulous generalities.

Drawings that are used conversationally, as an exchange among people, show how social the drawing process can be. Mutually acceptable drawing on top of another person's sketch can be a contributive act, with the result seen as a jointly-owned piece of work. This can be a critical step in solving complex tasks among disparate group members.

---

66 I keep going back to speaking, to English. You don't have to be a professional speaker for words to be useful in your life. It's the same thing with drawing. You don't have to be a professional artist to have drawing be useful..." -Scott Kim

66 In the best interactions, people start to interact with the picture and make changes. When there's a good collaboration going on with someone, you actually begin to form a joint drawing...that's when the sparks really fly and something new can be made." -Rob Semper

Drawings also allow conversations to focus on fine levels of detail. In discussing a hinge for a door, for example, you might talk for a long time without specifying in which direction the door would swing. But if you drew the hinge, it would quickly become obvious that this attribute needed to be specified. It would also highlight the location of the screw holes, the size of the screws needed, and other critical details.

Drawing Is a Social Activity

The social nature of drawing isn't surprising if we remember our childhoods or watch our children. Children constantly draw together. Often they work side by side, each on their own productions; but they also look at each other's work, commenting on it and copying whatever they see and like. Methods for drawing pass from one child to another: the bunny rabbit that one child draws soon shows up in another's drawing. Advanced techniques, such as shading, also spread as one child learns to master them and others mimic this progress.

Children also delight in the drawing abilities of others, especially when they see their own ideas being portrayed on paper.

"Draw me a cat," says the child sitting in your lap, and you sketch out a rounded face with slanted eyes, whiskers and a fluffy tail. "Make him grey…make him bigger…make him sad…" Often smiles of delight and surprise accompany these sessions, as the child watches his or her own desires manifest on paper. Later you may see the child duplicate some of your techniques in his or her own drawings.

As much as children will mimic technique, however, what drives their drawing is really the need to express themselves. They are interested in competency only as far as it enables them to make an idea understood. Not until or unless the concept of "talent" starts to creep into their awareness do they start to falter, to judge their works against the efforts of others. It's a sad fact that as we grow older, drawing for the sheer joy of it gets left behind with other child's play: we grow out of it rather than into it!

"Somebody once noted that no matter how cows arrange themselves on a hillside, they always look picturesque and beautifully composed. By something like the same token, children always seem to draw beautifully, to communicate directly and artistically the things they have on their minds. Later, when their teachers explain to them how inept and inadequate their drawing techniques are, they turn into unartistic and frustrated adults." Peter Becker

We can change this, though. Start saying "yes" when your children ask you to draw with them. Draw a picture of your current project and put it on your door at work. Include pictures in your term papers and your business reports to make your ideas clear and frame your discussions around these pictures. Develop an "unhesitating sketch response" to show others what you mean and enlist drawing in helping you to solve your problems.

The "I Can't Draw" Syndrome Is a Self-Fulfilling Prophecy

You may feel inadequate about your drawing skills, and not daring to embarrass yourself in front of family, colleagues, or teachers. Although you are curious about this "drawing as communicating" approach, you truly don't intend to make it part of your everyday life. After all, you're already good at what you do, why mess it up with your artistic incompetencies?

Many of us have thought this way for a long time; in fact it's one reason that drawing typically takes place only in art and design arenas. Most people simply stop drawing when they are about eight, the typical age when those with natural talent are identified and pointed towards art classes and those without these skills are motivated towards math, social studies, and the other "serious" disciplines.

Interestingly, nobody really wins from this bifurcation. Visually talented students are never expected to do much in the academic disciplines, and those without "natural talent" are never given any training in the visual arena.

In fact, perpetuating the "I'm an artist" and "I can't draw" distinction creates an artificially divided society, and it inhibits one large segment of that society from ever developing communication skills to represent their ideas. It is as though, by second grade, we were to decide that only the best readers could keep reading books, and all the other students would have to start

focusing only on arithmetic. This approach strips people of their basic literacy in visual forms, and it discourages visual communications and conversations because ultimately there are so few people who can participate.

A friend used to teach art in a small school in New England. He tells the story of a ten-year-old who asked him if her drawing was "right." When he replied that it was well done and then asked her what she was trying to accomplish, she completely froze. She had no way to respond to a teacher who didn't tell her what was right and wrong. She had lost her self-confidence, her belief in her own visual interpretation, and was bewildered when the teacher wouldn't provide her with one.

As long as we think we can't draw, then we won't draw. As long as we are denied training, as long as there is no acknowledgement of the importance of drawing, then we won't draw. And if we don't draw, we won't get any better at it. Nor will we learn how to take full advantage of our visual thinking capabilities.

---

66 If you look at kids' drawings, where they put the sun and how they start to treat shadows...having the sun with rays coming off of it to try and represent the warmth that you feel from the sun...kids actually play out their understanding of the world in the drawings that they do." -Rob Semper

66 I find the kids really appreciate the fact that I draw fairly well. If we're talking about a Tyrannosaurus Rex, I can get him up there. They all ooh and aah. And when we're doing art projects I can help them out by maybe starting a drawing on the board that they can try and copy..." -Alison Quoyeser

Knowing, Seeing, and Drawing: Three Intermingled Perspectives

If drawing has not been a part of your life, you may become frustrated when you start to draw again. You may encounter a number of problems that discourage you from staying with the process long enough to develop your skills.

Comparing Your Work with Professional Art

For one, you'll begin to notice how good most artwork around you seems. This is because you usually see only the finished drawings of professional artists. You don't have access to their discarded attempts or rough drafts, and so you have little sense of what it took to create these drawings. Finished artwork may also include a number of sophisticated techniques that are impossible to separate into meaningful units. It's a bit like trying to learn to play the violin by listening only to the performances of a major orchestra!

Giving Yourself Permission to Draw

You may find yourself under a good deal of social pressure to produce "good" visuals. Unless you are part of an active visual culture, or in early elementary education, you may not be granted the permission to do whatever it takes to explain your ideas. Until you get this permission – or decide to act without it – you and your colleagues will not get the practice you need in communicating with drawings. Sometimes this means changing the tenor of the culture in which you work. Encourage the use of informal sketches as pictorial conversations, and support each other's efforts to draw ideas, no matter how primitive or abstract the results.

Trying to Draw What You Know

Sometimes what you know will get in the way of what you draw. You may find it hard to draw a cup without its handle because you know the handle exists, even if you can't see it. You may want to draw both eyes in a person's face, even though you can see only one of them, because you know there are two eyes. It may be hard to draw a building smaller than a person, even though the building is in the background and the person is nearby, because you know buildings are bigger than people.

This tension between "what is" and "what appears" is a classic one, in life and in the development of drawing techniques. In the arts, spirited debates abound about whether or not it is the role of the artist to show "what is" or whether the magnificence of good art lies in how it portrays what the world looks like to the artist. For example, although the colors of Vincent Van Gogh's bedroom were probably not blue, orange, and green, these are the colors he chose to portray how the room appeared to him. People once thought this kind of impressionistic painting perverse and trivial, but many now acknowledge that the artist's reality is a valid part of great art.

66 You don't have to be a skilled artist in order to communicate these ideas. You just have to be able to draw a stick figure, which most people can do. One researcher drew this cute little boat, this research vessel that was the interface. It was real rough, but it worked and we could all see it." - Frank Wiley

In the context of visual thinking, your drawing has the same permission as Van Gogh's – to show your idea, your perspective, as it appears to you, the artist. Even so, you're going to need to address the basic tension between "what is" and "what appears," for if other people cannot understand your drawings, you won't be able to even begin your pictorial conversation.

Consider a Cube

You can do a simple exercise to illustrate the difference between "what is" and "what appears to be." Draw a cube in your sketchbook.

Now look at your drawing. How many right angles does your cube have? How many faces of the cube are showing? What is the cube's position relative to your point of view – above your line of vision? Below? Straight ahead? Is it centered, or slightly to the left or right of your position?

If you have drawn your cube the way we drew the *VizAbility* cube, you may have four right angles on the front face of the cube, but probably no more than these. Similarly, you've probably drawn the cube so that only two or three faces are showing – front, top, and right, for example. You may have drawn the cube as though it were slightly to your left or right (though you probably didn't think much about this when you made your drawing).

Next, consider everything you know about a cube. How many right angles does a cube have? How many faces? From which angle should a cube be drawn?

Intellectually, you know there are six faces to a cube. There are also 24 sets of perpendicular lines, four on each of the six faces. And there is no obvious place from which to draw a cube; it is a cube, no matter which of the infinite possible angles and viewing distances you choose to view it from.

If you know all these things about a cube, why didn't you show them in your drawing?

Chances are, you probably ignored the cube's principle features in order to draw it. By deciding to make a "realistic" drawing, you chose to view the cube's appearance at a particular point in time and space, and to render this.

Of course, there's a chance you went to the other extreme, showing the cube more like "it is" than as "it appears." Children will often draw a cube more like the unfolded cube that is used as the *CD-ROM* Main Menu, for example. Interestingly, this unfolded cube does have six faces and does have 24 right angles. It has all the properties of a cube, yet it is not what most of us think of as a drawing of a cube.

It's an example of "what you know" instead of "what you draw."

" Drawing lets you bring the world around you into your conversations." - Gayle Curtis

Dürer's Nude, or the View from Your Windshield

The artistic tension between "what is" and "what appears" was at the heart of the development of perspective drawing in the Renaissance. A famous woodblock by Albrecht Dürer, *Draughtsman Making a Perspective Drawing of a Woman* captures some of this tension.

In this woodblock, an artist is shown gazing at a woman reclining in front of him. Between the artist and the model is a see-through plane, which has a grid pattern on it. The artist is looking through the plane with one eye, from a fixed position. He is recreating the image of the woman on a piece of paper in front of him; the paper has the same grid on it as the see-through plane.

This technique creates a number of surprises. Principal among these is the degree of distortion. In Dürer's illustration, for example, the artist discovers as he draws that the legs and knees of the woman are emerging as extremely large and prominent. The woman's face, often important to a human figure, is only a minor part of the drawing. Her breasts, which some might argue are even more important, are completely left out in the drawing because they are occluded by her knees from that particular angle of view.

You can experience this phenomenon directly by paying attention to the view from your car windshield as you are driving. If you casually reflect on what you see, you'll notice such things as buildings and trees and people walking along the sides of the roadway. If you were asked to describe the prominent elements of the scene you would most likely mention these objects.

Now imagine a "freeze frame" of your view as you are driving down the same road. Imagine tracing this view on your windshield (or go ahead and take a quick photo). What would this tracing be like?

You would probably find that almost half of the image consisted of roadway. It would dominate your view as a large triangle whose base is the width of your windshield. The buildings, trees, and other things that you thought were "the main story" would actually be quite small and relatively unimportant. If you took a photo, you would find that you had a very ugly picture, with many of the details totally unrecognizable.

Actually, you may have encountered this predicament before, if you tried to take photographs of an exciting event. The lavishly costumed actors moving on an outdoor stage turn into tiny colored figures glimpsed between the large dark blobs of the people's heads in front of you. The rare bird flying into a bush turns into a smudge of feathers in a panorama of green leaves.

What does all this mean? It means that a literal rendering of scenes can make for very unsatisfying drawings, even when accurate. It means that you are ready to consider a range of techniques developed by artists and others over the years to portray "apparent realities."

Engineers and Artists

In the Renaissance, artists attempted to minimize the difference between "what appears" and "what is" by developing precise rules for recording both viewed and imagined scenes. One-, two- and three-point perspective systems were established; they dictated the ways in which spaces might be rendered. These well-delineated systems were logical, mathematical, and precise enough to be replicated by students, and consistent enough for teachers to judge a drawing's "accuracy."

These systems did not accurately reproduce reality – they just provided consistent techniques for rendering scenes so that they were interpreted as realistic.

A number of variations on these perspective systems have been developed for particular purposes. Some show the attributes of objects rather than their appearance, rendering displays that are more like the unfolded cube. Drawing systems like these, including orthographic projections and sectional analysis, are often used to communicate the details of an object that are designed by one person but built by someone else. Their value lies in the fact that accurate information about distance and form can be obtained from these renderings, information that is difficult to gather accurately from many perspective renderings. Provocatively, many computer-aided design systems (CAD) provide representations that both preserve accurate data and render a realistic appearance from any angle.

66 The vast majority of humans learn in a more meaningful way if they have something they can make their impact on, such as sketching on that white piece of paper. It's as if ideas locked up in the mind are suddenly brought to fruition on the page."
-Alison Quoyeser

Seeing and Showing What You See

You can choose to simply show what you see in a drawing, without any intent to make things appear realistic. You can develop an "appearance" drawing that focuses on what you see, not what seems realistic or what illustrates what you know about something.

Drawing just what you see can be extremely important in helping you to see the world around you, as noted in the previous chapter.

The interaction of seeing and drawing can also be quite helpful to you in developing your drawing abilities. By focusing on the perceptual aspects of the world rather than the conceptual aspects, you can begin to render subtleties of the visual world – the colors, the changing shadows, the details. Including these in your renderings will ultimately give your viewers a sense of what you see and will typically engage them in a serious analysis of your drawing.

Once you feel competent in these "appearance drawings," you are ready to take advantage of some of the techniques that artists have developed over the centuries. You can learn to use these "tricks" to help you create "apparently realistic" drawings – drawings that can represent your ideas, that are good enough to engage serious conversations.

Showing What You Imagine

Just as seeing and drawing interact, your imagination interacts with your drawing. For often you will want to show something in a drawing that either does not exist anywhere in the world, or that doesn't exist in your immediate vicinity, hence requiring that you draw it from memory rather than direct perception.

Drawing something that doesn't exist anywhere in the world is a very typical task of a designer or inventor, and is often the goal of these professions. The task is to make something new, and so drawings are designed to explain this new concept to others, whether it is a new town, or a new dress, or a new piece of playground equipment. Drawing something that doesn't exist is also the purview of dramatic professions, as children's book illustrators portray imaginary worlds that children enter into and as set designers suggest millions of moods and activities in drawing out their ideas for an opera or a play.

66 The first part I see is communication to yourself. You have these ideas that may be somewhat worked out... Until you get them down on paper, they remain kind of nebulous. Being able to put them down on paper for your own use is a necessary first step for most designers." -Denny Boyle

Oftentimes the drawings that arise in these situations are quite stereotypical. Typically, they intentionally don't show any particular person or place, for example, or any that might actually exist. Instead they show a combination of elements selected to communicate a particular perspective or to emphasize a particular aspect of something (e.g., the freckled nose of a little boy or the tropical flair of a location). Often these drawings will wildly misrepresent realities on purpose, to show unicorns and magic worlds, or future scenarios that show the impact of acid rain in one hundred years should we not address this ecological issue.

Stereotypical drawings also often result from drawing from memory, though often not intentionally. As you draw the entrance to your house or your mother's face or your childhood home, you will probably forget many details and render some in exaggerated form. You will simply forget some of what you have seen, and rely instead on your general understanding of the elements and your general repertoire of drawing techniques and logic.

As you show what you imagine you will probably refine your imagination. You will do this as you look at your own drawing, and as you receive feedback from colleagues and friends. As with seeing and drawing, the interaction of imagination and drawing can greatly enhance your own abilities and your performance.

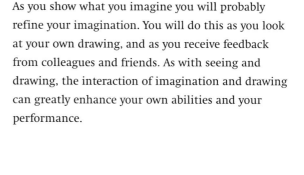

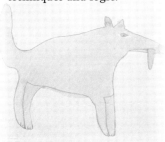

66 If I have the children draw very carefully, then I exercise their powers of observation...when we studied the human eye, the second graders drew eyes in all kinds of goofy ways...then I had them look at their neighbor's eyes and draw the eye accurately...they could really see it." -Alison Quoyeser

Techniques to Draw What You See and Imagine

MY HAND POSITION LENT ITSELF TO SEEING THE CONTOURS AND I FOUND MYSELF MUCH MORE INTRIGUED VISUALLY BY THE SURFACE OF MY HAND AS OPPOSED TO ITS OUTLINE

Your goal at the moment is to reclaim your drawing abilities, or – if you are already skilled in sketching – to refocus them a bit.

There are a number of very basic visual elements you can use to begin drawing with confidence; exercises to address each of these can be found in the corresponding sections of the *CD-ROM*.

Learning about each of these elements can launch you in your own drawing experiences. They can provide you with a foundation for participating in drawing-based collaborations. Don't spend much time understanding these techniques. Theory is not important just now. Your task is to begin to draw, to draw, and draw some more to develop your skills.

Drawing Basic Forms

Contours are the edges that define an object, that separate one object from another. If you learn to pay attention to these edges in the world around you, you'll be able to render the basic shape of an object, from different views and in different positions. It is an effective method to show what you see; it gets you drawing quite well quite quickly.

Studio art classes often contain a number of contour drawing exercises, designed to focus your attention on these edges and to give you experience in rendering what you see.

411 Contour Drawing Introduction
412 Draw Your Hand
413 Look at Other People's Drawing

Before you start drawing, try some Drawing *WarmUps* from the *CD-ROM*. Draw some simple shapes to loosen your arm, improve hand-eye coordination, and increase your confidence in drawing. 760

One contour drawing lesson is provided on the *CD-ROM*. Its purpose is not only to give you experience in drawing a readily available object – your hand – but to focus your energy on what you see rather than on what you draw. It is a tool for invoking a "perceptual shift," one that moves you away from symbolic representations ("I'm drawing this hand") and toward direct perception ("I'm experiencing this hand").

The Lesson

Ready yourself for drawing in your *Sketchbook*. Choose a clean page, giving yourself lots of room. Date the page so you can refer to it later, since you may want to try this more than once and compare this sketch with your later efforts.

Take your time with this lesson. Don't simply listen to all the instructions in a few minutes and consider yourself done. This is not a browsing exercise. It is a time to draw. If you take less than two minutes with this exercise you are not going to gain much from it.

Don't look at your drawing and don't even think about whether you are making a good or bad drawing. If you happen to lose track of where you are, just keep drawing. Draw what you see. You will have plenty of time to look at it later, and you will have many opportunities to do this exercise again and again.

Once you're familiar with this activity, change your method slightly. For example, look at your paper once as you are drawing. Or try looking at it more often.

Relax the requirement to draw one continuous line.

Choose different objects. Try your shoe, your arm, the items on your desk, the food on your plate.

The more you do this exercise, the better you'll become at drawing what you see, the more proficient you will be in showing the outer form of objects, and the more confident you will become in your seeing and drawing abilities.

Reflections

What did you think of your drawing once you completed it and looked at it? Did it surprise you? Did you like the drawing itself? What about the process of creating it?

Many people who do this exercise actually like the drawing that results, even though creating a product wasn't their goal, and even though they had relatively low expectations about the results. The drawing serves as a memory of the process, and this is what they like

Look carefully at an object – such as your hand – and trace its outline without looking at your paper. Slowly follow the contour of the object with your eyes while you are drawing. Use a single, continuous line.

about it. Others find the abstract quality of these drawings aesthetically pleasing or amusing.

The GALLERY section of CONTOUR DRAWING on the CD-ROM contains examples of other people's efforts. Take some time looking at these drawings. Although most hands are quite similar, the drawings vary widely due to the different positions of the hand, the details people chose to observe or ignore, and the sizes of the completed drawings.

Notice how some of the drawings are very constrained and "tight," and others are looser and "flowing." Which of these is most like the feeling you had while outlining your own hand? Which of these seems better to you as a final product?

It should be clear by looking at the examples that there is no "right" way to draw, only a range of approaches. Learn to be flexible, to master different approaches, so that you can choose the one that's most appropriate for a given situation.

Make it a habit to notice good ideas in other people's drawings in order to improve your drawing abilities. Most art or design courses display student work, and often the instructor will discuss all of it openly. There is a great advantage to looking at a large number of solutions to deepen your understanding of the same problem or task; at the same time you can face the problem itself and consider multiple solutions.

Foreground and Background

It's helpful to be aware of the outer edges, or contours, of objects when learning to show their form in a drawing. Many objects and scenes, however, are more visually complicated than your hand. Another way to assess the basic form of an object is to consider the "negative" or "non-object" spaces surrounding it.

Attending to negative space can be a good technique for viewing one set of objects in relation to another set, or in relation to their background. Like contour drawing, it is a basic technique that will "get you off the dime," beginning to draw anything quite easily.

The Lesson

Take some time experiencing the objects shown on the CD-ROM in BETWEEN SPACES. Then try drawing some of these objects; show the "positive" space by sketching out the contour of the negative spaces.

---

421 Between Spaces
422 Draw Some Negative Spaces
423 See Some Examples Others Have Drawn
424 Try Some More Scenes

Try drawing objects around you (or in pictures) by outlining the "in-between" spaces, the space between your fingers, behind the house, etc. Draw the non-object to show the object of interest.

This approach to drawing is rather non-intuitive, and you will need to persevere for a while before you see results. The shapes you render may seem meaningless, even if rendered accurately; but suddenly, the object you are drawing will emerge almost as if by magic. It will sort of "come out of the fog." Even though you have not focused on the object itself, your rendering of the negative space will have delineated it. By sketching what is "not the object" you will create its form.

Once you are comfortable with this approach, draw an object or two from your environment. (Some objects will be better than others.) Don't pay attention to the three-dimensionality of these items; look at them as though they were flat photographs.

Try drawing your hand using this method, afterward noticing how this drawing differs from the one you created using the contour drawing technique.

Make sure you don't get in the habit of "rushing through" these activities, just to get done. Instead, develop a habit of "lingering" over each drawing task, becoming immersed in the object and in your drawing of it. This ability to become absorbed in the task is a large part of the drawing process, and is critical to learning to draw.

Many drawings made with this technique are pleasing to the eye. They capture the "essence" of objects and scenes and delight our visual sensibilities.

What do you think of your drawings? Do they represent the objects and scenes well? What about the process of drawing the objects? Did you find it hard to focus only on the edges of the objects and to ignore all the surface details? Are you surprised at how much the drawings look like the objects without these details?

Light and Shadow

CONTOUR DRAWING and BETWEEN SPACES both offer techniques for drawing the basic outline of an object – or seeing it in relation to the space surrounding it.

Now it's time to consider another major element of the world around you, and its importance in the drawing process – the element of light.

To the naive eye the world is basically light or dark; that is, it is either bright enough to see or too dark to see. With a little more experience, one perceives subtle gradations of light and darkness. You play with the brightness knob on a television and learn how the visual quality in a scene changes according to the degree of brightness. You discover that sometimes it is too dark to take a photograph, even though it is bright enough to see clearly. In most cases, these simple understandings about light are sufficient for getting by adequately in the world or for haphazardly recording it on film.

However, this naiveté must be overcome if you want to develop your seeing ability, draw realistic-looking scenes and objects, or become a creative photographer. For the play of light is among the most defining of characteristics in our world. It reveals information about an object's size, depth, and texture. It defines an object's relationship to other objects. To ignore light is to ignore many of the basic elements of the world around you.

Pay attention to the natural light around you. Notice the shadows on buildings, and how they change during the day. Watch the light shifting across flowers, highlighting different elements as it changes. Look at the differences in color that result – a red in the shade is a very different color from one in bright sunlight.

Notice the morning's light and compare it to that of evening. Fine tune your awareness of mid-morning, midday, and mid-afternoon light. Take the time to watch as light gradually shifts across an object. Or check an object only once or twice a day, and notice the large contrasts in lighting that are not evident on a continuing basis.

Do you notice the difference between the quality of the light today and three months ago? If not, pay some attention over the next few seasons. Notice how light falls on hilltops and in gulleys, across great open spaces and within enclosed areas. Observe the light through a window facing north, then one facing south. Get a sense of why artists take great care in choosing particular locations for their studios.

Use artificial sources of light to get different effects and observe the results. Try changing the lighting in your room, bouncing it off the ceiling, or highlighting particular areas while downplaying others. Notice how this changes the dimensionality of the room, and the "feeling" of the objects it contains. Stage set designers, photographers, interior decorators, and others use a heightened awareness of light to create desired effects. They have perfected the art of using light to influence how others see things.

As children, we used a flashlight to illuminate our faces from below, making all kinds of terrible grimaces. The effect was truly horrific! Try lighting your own face (or a friend's) this way, moving the flashlight above, below, and to the side – and see how your appearance changes from happy, to sad, to menacing, all because of different lighting.

There is a light performance going on every day in the world. Now that you are developing your visual abilities it is time to notice and enjoy this – particularly as you learn to draw. You'll realize that you need some basic techniques for showing light and shadow. Shading creates the illusion of mass and volume in the objects you render.

Simple Light Sources

In the natural world there are a myriad of both direct and indirect light sources, as light from the sun bounces on and between different surfaces. In the world of artificial lighting, which is deliberately set up for particular effects, there can be either multiple or single light sources.

The Shading Exercise

In SHADING EXERCISE you can explore how simple light sources fall on different sets of objects. Your task is to select a light source, then "shade" each object until it portrays the combination of brightness and shadow

appropriate for that light source. You shade an object by clicking on its surface(s). Remember that all the light sources in this activity are at a 45° angle – that is, slightly above the objects. A chime will sound when the shading is appropriate.

Try using all or most of the light sources for each object or set of objects. As you become familiar with an object, you'll get better at shading its surfaces.

Although you are not actually drawing in this activity, you are still developing "an eye" for the properties of light and shadow, and you will be able to transfer this intuitive knowledge to objects in the real world. For if you can gain an intuitive sense of shading, you can use shading to give a sense of volume and mass to your views that will be convincing.

A Matching Exercise

In the next exercise, MATCH, you choose the scene that has been shaded appropriately for each given light source. Your task here is to notice subtle distinctions, not only among different sets of objects but within the same set – that is, the way light is reflected on the apple may be correct, but the shadow it is casting may be wrong. As you become comfortable with this task, try it backwards; look at the alternatives and try to determine where the light source is for each.

Place one or more objects beneath a single light source, such as a lamp. Sketch them, paying special attention to the range of light values, from deep shade to highlights. Reposition the objects so that light strikes them from a different angle and draw them again.

Reflections

There are two categories of skills you can apply to these shading tasks. You can rely primarily on your reasoning abilities ("If the light is to the left then it would shine brightly on the front of the cup, which means that the ball would be in the darkness"). Or you can be more direct, and rely upon your immediate perceptual judgments ("This looks like I would expect from this point of view; the shading and shadows must be correct").

In one case you are relying on logic-based problem-solving capabilities; in the other you are using a more intuitive, sensory process. Both of these capabilities are important, but in drawing you need to rely more and more on perceptual skills. You can use your analytic skills to refine your drawings, but you should use your visual imagination and intuitions to begin the process.

This type of visual "instinct" allows you to relax and rely on your intuitive sense of form, shading, and perspective in order to render something. Any important inaccuracies can be refined through your analytical skills later. Remember your task in drawing to communicate is not really to be completely accurate in most cases. It is to render objects so that they appear realistic so as to engender good conversations.

Learning How to Shade

The MATCH EXERCISE not only develops your sense of light and shadow, it also offers examples of how you might indicate shading on different objects. Notice how the artist used cross-hatching, for example; this is a basic technique for building up areas of shadow. Or notice how diffuse shading, which ranges from light to dark, is indicated on different surfaces. Note how simple shading on the edges of the objects can lend a three-dimensional effect; how line weight is used to make visual statements about depth and relationship; and how contrasting shading on adjacent surfaces conveys important information.

Try drawing some of these scenes in your *Sketchbook*. Make a quick sketch of a scene in the MATCH EXERCISE without shadows. Make multiple copies of this sketch. Now, choose a light source and shade your sketch appropriately (without looking at the *CD-ROM* drawings!). Compare your sketch with the correct answer on the screen and refine your drawing. Try this a number of times, using light sources from different angles, to transfer your intuitive sense of shading to your developing repertoire of drawing techniques.

Draw your own overlay lines – to mark the horizon, the lines of convergence, and any vanishing points – on magazine photos.

Look for Other Examples

The MADE IN THE SHADE GALLERY has many excellent examples of natural light and shadow. Try sketching some of these images to enhance your awareness of the world around you. Use different tools – hard or soft pencils, charcoal, or pen – to experience how different media create shadows. Charcoal, for example, is very soft and can create a dark shadow very fast. A lead pencil will never give you the same sort of depth, but will allow you to sketch in basic indications and build up dark areas gradually, often by switching from a harder to a softer lead. Pen works well for cross-hatching, but it is a little harsh for gradual variations in shading.

Create your own still life by setting up an arrangement of some objects from your environment. Use a lamp to illuminate them. Draw the objects, including the shadows. (If you wish, use the contour drawing technique to establish an object's basic form before trying to shade it.) Move your lamp around the still life to vary its shading. Try moving the lamp a few times without drawing, just looking at the changes. Remember, you draw first with your eyes, then your mind, then your hand.

As you gain confidence in your perception and drawing ability, change your position as well as the lamp's, viewing and drawing the objects from slightly different angles. Finally, when you are comfortable with this exercise, try adding another light source. Notice the subtle and complex effects this has on your objects. Try drawing these effects.

Three-Dimensions in Two: Perspective

Like the more general problem of how to depict the world realistically, the puzzle of how to represent a three-dimensional world on a two-dimensional piece of paper has fascinated people for centuries.

At an extreme, there is the hotly-debated approach of representing the world "just as it is," so that viewers are unable to tell the difference between real objects and representations of these objects. An example would be tromp l'oeil, where a realistically rendered painting is superimposed upon a real object – such as a painting of window shutters and a window box full of flowers that surround a real window. The use of a real object reinforces the illusion created by the painting.

A more modern example is the current attempt to create "virtual reality" with computers. These programs recreate the feeling of "being there" to such an extent that one is unable to distinguish the fabricated "reality" from the true world.

On a large sheet of paper, use sweeping motions to draw lines between widely spaced dots. See how smoothly you can make these gestures, and how accurately you can gauge the straightness of line needed to get from point A to B. 760 Random Lines

Fortunately, our drawing goals are more conventional and less difficult to achieve. We are concerned with creating the illusions of dimensionality and depth in more traditional ways, to use simple perspective to bring ideas to the table in a convincing way.

Making Spaces Appear Three-Dimensional

In medieval European art, spatial relationships were typically indicated by the position of objects on the page. Items that were farther away were drawn smaller and placed at the top of the image; nearby items were drawn larger and placed at the bottom.

Oriental artists also used this convention, but heightened the illusion of space and depth by leaving blank, misty regions between the foregrounds and backgrounds of their paintings.

As we discussed earlier in this chapter, Renaissance artists attempted to represent spatial relationships in more realistic ways that mimic our real-world visual experiences. Currently there are a number of very complex perspective rendering schemes, many of them implemented on CAD systems.

From the Renaissance to the present, perspective rendering schemes have included two important attributes. The first is they are repeatable; this means that there is an accepted perspective for any given object in each system, allowing an accurate interpretation of a physical space, free of subjective judgments. The second is that these systems make the world appear three-dimensional to viewers.

In professional arenas, the first of these attributes is extremely important, as it lets one person use another's drawings with confidence. In the context of visual communication, however, it is the second that is the most critical, and thus it is this element that is our focus in *VizAbility*. Our goal is to touch upon a few basic techniques that will allow you to draw objects and their spatial relationships in ways that encourage viewers to interpret diagrams accurately and to infer this dimensionality in more realistic ways. As always, the more clearly you can draw your ideas, the more effectively they will be interpreted by others.

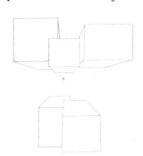 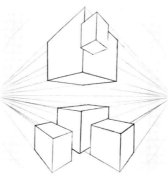 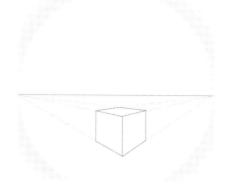

66 Using a steady, sweeping motion, throw a line from the top of the cube's edge to the vanishing point on the right side..."
443

One-Point, Two-Point and Three-Point Perspective

The three basic perspective systems that are widely used (one-, two- and three-point perspective) each provides a set of conventions for rendering objects in space and is helpful for achieving different things.

In many ways, one-point perspective is the simplest. It is based on the general observation that when you look at any set of parallel lines (such as a road) going away from you, they will appear to meet, or converge, at the horizon.

"In one point-perspective, the front face of a cube looks square and the other edges seem to converge at a single point called the vanishing point, on the horizon. This type of perspective is good for showing wide open spaces, and the insides of things." **441**

Two-point perspective is a bit more complex. This system is based on the observation that objects with a single vertical edge will converge at the horizon on both the right and the left side.

"In two-point perspective, the vertical edges stay vertical, and the other edges seem to converge toward two points on the horizon. This perspective is good for showing things that have two interesting sides." **441**

A two-point perspective is used more frequently than a one-point perspective system, as it is more generally applicable to complex scenes and results in drawings that look more realistic.

Three-point perspective is quite a bit more specialized and not as easily recognized in scenes around us. Knowing how to render this perspective is extremely useful in fields such as architecture, however, where one wants to maintain the appearance of parallel lines within three-dimensional spaces.

"In three-point perspective, one corner points towards you, and the edges seem to converge at three points on a big circle. This circle represents your field or 'cone' of vision. Three-point perspective is good for showing tall buildings or exaggerating the size of things to make a special point." **441**

Each of these perspective systems assumes a stationary point of view. It also assumes that the scene is projected upon a plane. It is important to acknowledge that none of these constraints is applicable to everyday seeing.

Vanishing Points

Look at the world around you to find models that motivated the design of perspective systems. Look particularly for convergence of parallel lines in the distance, at the horizon. Notably, these lines converge at a single point, the "vanishing point."

You can train yourself to look for vanishing points by reviewing the images in the VANISHING POINTS activity. Look at these images carefully, identifying the vanishing point(s) and horizon for each scene.

The orange overlay lines represent the converging of parallel elements. The yellow line represents the horizon. A black circle indicates a vanishing point.

It's worth noting that man-made creations, not natural scenes, best exemplify one-point perspective. Many of these, such as temples and formal gardens, were deliberately constructed to focus the viewer's eye on predominant elements.

VANISHING POINTS also includes images that represent two- and three-point perspective. In these images, the vanishing points are typically off the screen, and the construction of the perspective is not as clear to the naive viewer.

Throw Lines

THROW LINES offers some techniques for drawing a cube in two-point perspective.

Watch and listen as a cube is constructed in two-point perspective. Learning to "throw" your lines is critical in this exercise. It accomplishes two things: First, the steady, sweeping motion of your arm from one point to another improves your hand-eye coordination and results in a straighter, more accurate line. Second, this technique gives you experience in loosening up when you draw; learning to "unclench" your arm and hand will result in fluid, more "alive" renderings.

Try the THROW LINES exercise a number of times, each time drawing the leading edge of your cube in a different location and at a different size. Become comfortable with the sensation of throwing your lines, and notice how much smoother and more accurate they become.

Reflections

Use this technique to draw a few cubes in a scene. If you feel daring, try drawing a few different objects in this perspective space, for example a rectangle or a sphere. Experiment with non-parallel elements in this space.

Notice that as you draw objects near the edge of your cone (or circle) of vision, they appear distorted. This does not mean you have followed the rules of this system incorrectly; it's simply a property of this method of drawing. It does mean that you should select this method: (a) when you are working within a large cone of vision, and (b) when most objects appear in the central area. Remember that this perspective drawing technique, like others, is simply part of a set of conventions for rendering spaces. It is not necessarily the only way to show "reality," and you should choose among the alternative systems available in order to best match your goals.

Square Faces

In Throw Lines, you simply guessed at the location of the two vertical lines that defined the sides of your cube. Unless you were viewing the cube in the center of your circle of vision, the length of these lines and their distance from the leading edge of your cube would not match. The angle from which you view a cube affects how the left and right faces should be drawn.

If a cube is drawn to the right of center, the left face should appear larger, and the vertical line that represents its outer edge should be longer and farther from the cube's leading edge. The right face should appear smaller, and the vertical line that represents its outer edge should be shorter and closer to the cube's leading edge.

For a cube drawn to the left of center, of course, the exact opposite is true.

The Square Faces exercise gives you a chance to experience this phenomenon and develop an intuitive sense of proportion within a two-point perspective space. Your task in this activity is to correctly place the side edges of a cube, given a particular circle of vision, horizon line, and leading edge.

Try the exercise. See how quickly you can make judgements and how many examples you can get right. (When you complete the drawing accurately by moving the L and R indicators to the correct position, a chime will sound and the rest of the cube will be shown.) Do this again and again until you don't have to think much about it, and you will able to see quickly and naturally what's "accurate" in this perspective space.

Reflections

Now that you are developing a certain intuitive understanding about a cube drawn in two-point perspective, try drawing cubes again in different areas within the circle of vision. Draw the whole cube, showing both obscured and visible faces, as if it were transparent.

Don't just practice drawing the cube; draw in all of the converging lines as well, noticing how this framework defines the cube.

As you master the skill of drawing an accurate cube in perspective, you'll find that it can serve you in many pictorial conversations. You now have building blocks for creating other objects and shapes. In and of itself, a cube is also a handy device for representing many items. For example, you might sketch a set of cubes in different positions to represent a city; or to show the location of containers at a docking facility; or to symbolize the elements of a museum exhibit.

Drop Cubes

Drawing a cube in a particular location requires that you draw the edges and sides in a particular way. This is what you have been practicing in our previous exercises. Conversely, a cube with a particular shape provides you with information about where that cube is located within the circle of vision. DROP CUBES will now give you an opportunity to practice working with cubes in this manner. Its goal is to further your intuitive understanding of how objects sit within a perspective space.

Your task is to place a cube correctly without the aid of vanishing points or converging lines.

Try it. Drag each cube that appears on the screen into its correct location. Keep track of your accuracy if it motivates you in developing an intuitive sense of this space. Start over a number of times and note how your ability to recognize where the cube belongs progresses.

Basic Component Shapes

Now that you have gained some experience with recognizing, constructing, and placing cubes in a perspective space, you can begin to use them as foundational elements for more complex objects. DRAW ALONG shows you how to draw a coffee mug and a chair, each derived from basic cubes.

A Coffee Mug

Using a cube as your framework, you can easily construct a cylinder, which can serve as the basis for many objects – in this case, a simple coffee mug. Watch the movie and then try sketching along with the step-by-step instructions. Take your time doing this, allowing at least 15 minutes. Try it again and again, until you feel confident drawing both the cylinder and the mug. Try drawing the mug in different locations, using

451 Introduction to Draw Along
452 Draw a Coffee Mug
453 Draw a Chair

66 To draw a coffee mug...begin by drawing an ellipse. Draw with a loose and easy movement, making sure the ellipse touches all four faces of the cube's top face." 452

your experience with cubes in different perspectives. Also try drawing other cylinder-based objects such as soup cans, silos, paper towel rolls, and so on.

A Wooden Chair

Again, you can begin with your familiar cube to create quite a handsome chair. You can use your experience in drawing throw lines to create the back of this chair in proper proportions. You can use your contour drawing experience to elaborate on the chair's details, and your basic experience in shading to accent particular parts of this chair. As with the mug, take your time drawing. Try it a number of times to increase your confidence.

Reflections

The DRAW ALONG exercise lets you put together the basic lessons of this section to produce a detailed drawing of a real object. You are beginning to learn how basic shapes can serve as the framework for more complex objects. Learn to see these primitive visual elements in the objects that surround you. How many cylinders and cubes, for example, form the basic shape

of a motorcycle? A tractor? A food processor? What about natural scenes, such as a garden, a house, or a landscape?

Perhaps the DRAW ALONG section will launch you on a sketching spree, and encourage you to try drawing everything that is around you. As you draw you will master the basic techniques already introduced to you – contour drawing, negative space, shading, perspective, and basic component shapes. For these are all complementary. Each will be particularly useful and appropriate in different situations. Some of your drawing experiences will hopefully prompt you to learn new techniques, ones that extend beyond those covered here. One way to extend your experience is to begin to notice how others draw. You can learn a great deal from such observations as well as from other resources – people, drawing classes and materials. Do whatever you can enhance your drawing abilities and knowledge.

Whether you engage in it for the purpose of art or communication, for personal expression or the capturing of ideas, take pleasure in your drawing. Re-experience the sheer joy of a child interacting with paper and tools. Refresh your seeing, your coordination, and your deep instinctive knowledge of how things appear. By doing so, you will give yourself an invaluable means of enhancing the way you experience the world, and for representing your ideas about it.

> 66 To draw a simple chair, start with a cube in two point perspective...draw two lines parallel to the top face of the cube; these will form the seat of the chair..." 453

Beginning to Draw: Putting It All Together

You have now been given a basic set of tools for developing your drawing abilities. You are mastering the skills you will need to bring what you see in the world into conversation.

As you feel more comfortable about your skills, you won't have to focus so much on shading or contour or perspective. Instead, you'll be able to think about the idea behind the object and to call on the appropriate means to represent it.

One wonderful thing about drawing is that a practice field is always available. Most real-life situations offer great sets of objects to draw.

DRAWING OBJECTS gives you three such everyday examples to practice on. These are the same objects you encountered in SEEING OBJECTS. As with SEEING OBJECTS, the tea pot, milk carton, and pen are all navigable movies; that is, they can be manipulated by holding the mouse down over the object's image and "dragging" it right, left, up, or down.

Choose a particular point of view for one of these objects and draw what you see in your *Sketchbook*.

Spend at least ten minutes drawing each object. First try a quick sketch, then do a more detailed rendering. Select a different point of view and draw the object from this angle. Try different techniques, such as contour drawing or positive/negative space. Notice how your experience of the object differs with various approaches. You might also want to try different materials in your drawing: pencil, pen, charcoal, colored chalk, markers, etc.

“ What I wanted to get was that nice swelling shape... I found that the light was playing all over it, so I got a little bit lost on the shading... I wanted to show the crispness of the dark lines around the lid...the texture in the woven handle."
- Alison Quoyeser's Teapot 463

Learning How Others Draw

One of the advantages of DRAWING OBJECTS is that it gives you an easy opportunity to compare your drawings with those of other people. Of course, you can also do this by having your friends draw the same set of objects and then compare drawings, and you should – drawing can be a very social activity, both in learning about technique and in communicating ideas. To get started in this collaborative mode, you can listen to different people describe their drawings of each object on the *CD-ROM*.

These people drew exactly the same three objects that you drew. They did quick sketches, spending only a few minutes on each object. The only difference is that they viewed these objects in the real-world, whereas you drew them from images on the computer screen.

You had the advantage of seeing the objects as two-dimensional images. On the other hand, you had the disadvantage of not having a natural light source. Despite these differences, there should be sufficient commonality of experience to enable you to learn from these drawings and their creators' comments.

The Teapot: Rendering Perceptual Details

You may remember that when we asked people to describe this teapot in SEEING OBJECTS, their descriptions included many metaphoric and associative links. One person was reminded of friends and good times, another person recalled his experience with heat transfer, and yet another found no place for a teapot within his cultural norms.

In sharp contrast to their "seeing" descriptions, these same people describe their drawing experiences in terms of perceptual details.

Jerry says that he always selects one detail to emphasize when he draws; in this case, it's the handle. Alison describes how the changes in shading on the teapot confused her. Rob acknowledges his difficulties in identifying and rendering the compound shapes that make up the teapot. Denny is pleased to capture the basic shape of the teapot, what Alison calls the "nice swelling shape." Nicholas wants to capture the pointed spout, depicting it as extremely pointed in his drawing. Erika provides us with a basic drawing lesson, describing how to put together the pot's basic shapes and how to add a highlight to indicate its "shine."

> 66 It was a challenge because there's hardly a straight line in the whole piece of work...it was fun, too, because I got to be really free...all the reflections were kind of a challenge, to fake a little bit of those..." -Denny Boyle's Teapot 463

> 66 ... In order for me to be able to draw something, I need to focus in on one thing that gets me started. With the teapot, for me that...was the arc and the texture and the cross-hatching of the handle." -Jerry Slick's Teapot 463

The drawings are as diverse as the people who drew them. How each person drew the teapot seems to reveal how each person saw the teapot. Consequently, the drawings tell us even more about these personal views than they do about the teapot's actual appearance. They also give us a brief snapshot of the kind of drawing techniques with which each person is familiar – or at least the ones he or she chose for this task. This gives us a sense of the visual variations that are possible for a given object.

Are these inconsistencies wrong? Recall our discussion of "what is" and "what appears." If one considers drawing to be a mirror of the world, inconsistencies such as these would constitute a problem, and the task would be to train people in techniques that allow them to show things "accurately" and "consistently." Indeed, for many professions in which details are important, this is the required approach.

However, if we consider drawings to represent people's ideas, to be an illustration of their thoughts or an elaboration of their plans, then this variability is appropriate. For the viewer of a drawing, how someone shows something is as informative as what they show. For, especially in its early stages, this is what communication is all about.

How does your teapot drawing compare with these six drawings? What do you think it communicates? Would you like to change it in any way?

You will probably be able to gain some insight into technique by looking at these other drawings. More importantly, perhaps you can acquire a sense of how your drawing compares to the drawings of others, or drawings done "the right way" (whatever that is!). Perhaps you will take away a greater sense of permission to draw what you see, not necessarily what others see.

The Milk Carton: Fun Geometries and a Spout

People enjoyed drawing the milk carton; they liked its basic rectangular form and were amused by the cow on the label (she is the basis for a local advertising campaign composed of terrible puns like "Moo-tiny on the Bounty.") The spout was a challenge or problem for everyone, and there were a diverse number of solutions proposed for representing this element. The text on the outside of the carton was considered either irrelevant or central, depending on the individual's perspective.

Again, the drawings are quite varied. Alison's is particularly pleasing and well articulated; her architectural background becomes clear in this form as does her enjoyment of drawing. Erika's and Nicholas's cartons show the whimsy of this form; children often seem able to get to the essence of an object and show it with some character. Frank seems

66 This pen...it had some very nice elemental shapes to it, which was kind of fun to discover while I was drawing it...just these kinds of little curves that mark the ends, and the straightness of the barrel...the shadow had a dramatic effect on what I was doing." -Rob Semper 465

to have turned his carton into a child's dollhouse in his search to make it meaningful. Denny and Rob are captured by the geometric regularity of the object and are quite uninterested in the text elements on its surface.

How did you do with the spout in your drawing? Did you use one of the techniques employed by the six other artists, or did you develop your own? What about the text? Did you include this or not? Much of what makes drawing personal involves the selection of elements. Although everyone drawing the milk carton saw the text, most of them did not deem it important. You will make judgements like this all the time in your drawings; it's crucial to acknowledge this evaluation process so that you will be sure to emphasize what you consider important in a drawing.

The Pen: Making Something of Nothing

Almost everyone had a hard time finding anything interesting in the pen. Yet this seemed to encourage a careful examination of the object, and an attempt to discover a set of details worth describing. The clip, the hole, the barrel of the pen, and the way it was shaded took on more significant import than would usually be the case. Because these were the only elements to build on, each artist incorporated them to give at least some interest to the pen. As a result, the drawings are by and large more interesting than the pen itself!

How does your pen look? Were you able to make it interesting? Did you use the same techniques as the others, or did you invent some yourself?

Practice, Practice, and Practice

So you have put all your techniques together to draw these objects. What else is there to do?

Drawing, like any other endeavor, requires practice and attention if you expect to improve. Even if your current level of ability is well-suited for most of the work you do, and enables you to have the conversations you need to have, you must keep practicing. You need to cultivate your drawing abilities continually to keep them fresh and accessible.

Or, you may have mastered your basic techniques but feel reluctant about drawing with others—much like people who can speak Spanish along with language tapes, but who are hesitant about talking in Spanish class, never mind speaking to someone from Mexico! Like the budding linguist, you need to get over this hurdle. For drawing will not become part of your general language repertoire unless you freely use it with others as well as on your own. Don't worry about failing; worry about not trying.

It may also be that this basic introduction to drawing has piqued your curiosity and you want to learn more. You may find drawing such a pleasant, rewarding, and calming activity that you want to get better and better at it. Go for it! Find other resources to improve your abilities. Practice whenever possible. You may move well beyond basic communication and decide to become a professional in art or design.

> 66 The only thing in the pen that really interested me was that little hole that's kind of up on the top. If it hadn't been for that little hole, I don't think I would have been able to draw the pen at all." -Jerry Slick 465

Alternative Ways to Gather Imagery

The underlying assumption made here has been that drawing is the process of sketching on paper with a pencil or pen, and that the product is a picture on this piece of paper.

But there are other ways to bring the world "to the table" for conversation and analysis. Drawings can be more than still frames made with pencils.

Photographs are one example. You can take pictures of the objects you want to talk about – the window seat detail; the flowers for the garden; the equipment for the chemistry demonstration – and bring them to a meeting. You can use sets of photographs to represent different views of an object or scene, and to help you consider alternate opportunities.

Instant cameras make it especially easy to incorporate photography into a conversation. With the advent of one-hour processing labs, even standard photography can be very facile in ongoing conversations and meetings.

Hi-8 cameras are also becoming easier to use for gathering visual information. A number of manufacturers now make video cameras that include enlarged monitors for viewing either before or after you shoot a scene.

Graphic artists often rely upon "clip art" or stock photography in making visual presentations, rather than customized drawings. These convenient resources are becoming increasingly accessible especially as image database systems and on-line services become more common.

These alternatives are not substitutes for drawing, but they offer additional means of gathering imagery for conversations quickly and easily within casual contexts. They also broaden the notion of "real-world imagery" from a single still to a set of images, sequences, and movies. These alternatives broaden your options, allowing you to select and modify an image for your own purposes when personally creating a custom-made drawing is not possible or practical.

Nonetheless, drawings are still typically the preferred means of visual communication. On a purely practical level, they are more conducive to visual thinking; a pencil and paper is often available when more elaborate equipment is not.

Even this basic assumption about drawing is changing, however, as our tools become digital.

The Digital World Encourages Image-Rich Communication

Computer-based representations are fundamentally indifferent to whether something is in text or in images. Past constraints of processing speeds and memory, which kept images secondary to text and numbers, are being overcome.

66 Unless you can focus on the creative energies behind the technical stuff, it's worthless." -Frank Wiley

New categories of tools are being developed that make the use of imagery in communication quite spontaneous. New digital cameras allow images to be incorporated immediately into the computer environment for manipulation. Movies taken with videocameras can be digitized into computers. Fiberoptics allow image-rich data to be sent easily from one place to another.

Computer graphics are also becoming more accessible, moving from complex professional tools to more affordable casual tools. These displays can prove extremely useful in bringing visual plans into a conversation, especially when real-world references do not yet exist. Many of these programs offer different types of visual transformations, such as the capability of changing the point of view or the shading. They offer a powerful new form of interactive imagery for communications.

In addition to the burgeoning wealth of software, the machines that handle these digital representations are also becoming more portable and affordable.

These are some of the exciting opportunities on the horizon that will extend the use of drawing in communication and add many categories of representation and technique to those provided by pencils and paper.

The challenge will be to stay focused on what you are trying to show – for one can become transfixed by the innovative capabilities of new tools and techniques, and forget what one has to say.

Explore these tools and learn what they offer. At the same time, keep drawing in your *Sketchbook*, training yourself to see and show what you see. These basic skills will transfer to the new technologies and will be critical to your effective use of them. You'll discover that you need to have a basic understanding of perspective, whether the lines are drawn by you or a computer system; that you need to understand the contour of a form, whether you draw it or scan it into your computer; and that you need to know about light and shadow, whether you are going to sketch, take photographs, or select from software display options.

From Objects to Abstractions

What if you want to show someone an abstract idea, one that isn't "object-based"? What if you want to communicate about process instead of product? What if it's just too early to discuss particular elements in the world? Can sketching provide an approach to this as well?

Yes! Let's roll the *VizAbility* cube again and shift our attention to a different point of view. While drawing is a way to bring real world objects "to the table" for analysis, diagramming is a type of sketching that brings abstract or evolving concepts into conversations.

66 I think the best advice you can give someone who is interested in multimedia is to make something. You can read all the books in the world, go to all the classes, talk to all the geniuses...but you'll never really understand unless you make something. It can be simple, but just make it." - Frank Wiley

Big
argument

Diagramming

Diagramming is a way to capture ideas and make them visible. It is a skill...

to show relationships and to emphasize critical points.

Diagrams can act as a form of visual note taking or as a pictorial conversation

between ourselves and a colleague...

to which we attach divergent ideas, or a more public

presentation of public data.

Use the basics of this craft and see how this visual ability enhances your own work. **501 Introduction**

Introduction

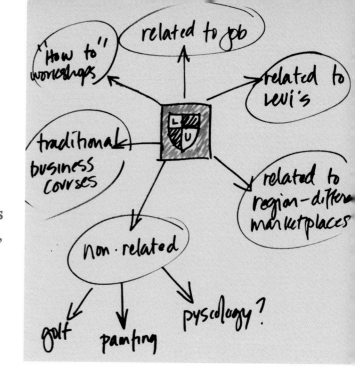

Diagramming lets you show ideas that are not obviously visual or visible. It lets you bring your ideas into conversations, and to provide records of their development.

Drawing lets you show the tree in your garden, or the chair at your dining room table. Diagramming lets you show what you think about the trees and the context in which you are thinking of the chair.

Drawing lets you show the detail on the door hinge that you are designing, or the set you are constructing for a play, or the flower you saw on your visit to Singapore. Diagramming lets you show interconnections between these elements, and their importance from your point of view.

Diagramming is the process of giving form to ideas that are abstract, ideas that may or may not result in "objects in the world," ideas that are in their earliest form and lack detail. While drawing is directed toward rendering objects that exist or are planned "in the world," diagramming is directed more to the realm "of the mind."

Diagrams can be used to record a process, an event, a structure, or a body of information. They can be composed of words, images, abstract symbols, or a combination of all three. They can be used in one-on-one or in group situations. In short, diagrams are an incredibly flexible way to illustrate, communicate, and record your ideas.

You can use a diagram to show the hierarchical structure of your company, or to explain the range of products you offer. You can sketch out the flow of a class lesson or the circulation pattern of a new museum. Diagrams can show sets of events over time or space.

66 Most people who are strong visualizers do model making, drawing, playing around, moving objects, creating displays, as a way of training their inner imagination. So diagramming and drawing is actually a fundamental way of thinking... Doing a diagram is a way of thinking." -David Sibbet

There are a number of ways in which diagrams can be useful. They can:

- Highlight ideas
- Emphasize relationships
- Make comparisons explicit
- Display different levels of analysis
- Reveal structure
- Group issues
- Encourage collaboration
- Reinforce your memory

Few people seem to diagram, but those who do tend to diagram a lot. It's as though it were a secret language, used only for special situations. It can become a major communication form once someone is converted to its use or it can remain a hidden language. For example, I have drawn diagrams of each of the chapters in this book, which probably no one else will ever see. These diagrams gave me a framework for my ideas, while working on the manuscript, a temporary scaffolding for ideas that will assume different forms as they evolve.

Although you may not be trained to draw diagrams, you probably know a bit about reading them.

We Live in a World of Diagrams

Diagrams can be purely personal arrangements, comprehensible only to their creator, or they can be formal, public structures.

One common but surmountable problem with personal diagrams is that people are often reluctant to share their thoughts before their ideas are completely developed. We're often uncomfortable with others'

rough, squiggly lines, and we may wonder just what these squiggles have to do with the magnificent and important concepts we're trying to resolve. We find the ambiguity and the roughness of personal diagrams unsettling.

In contrast, most of us have had years of experience with public diagrams and find these useful. Good communicators – including teachers, salespeople, transportation system designers, and anatomy book illustrators – have been using diagrams effectively for a long time. In doing so they have developed many inventive ways to make complex ideas visible to their audiences.

The Paris, London, and New York subways all use public diagrams to show how to navigate through their systems, and millions of people read them every day. The weather map on your television or in your newspaper is a diagram around which many of us make our daily plans. Bus and train schedules are diagrams that organize complex schedules in easy-to-read formats (most of the time!). Musical scores are a form of visual diagram used by musicians. Globes, the periodic table, maps, menus, multiplication tables, blueprints, schematics, family trees – to name just a few – are all examples of great public diagrams.

Several well-known books contain superb examples of complex ideas presented as diagrams. Harry Robin's *The Scientific Image: From Cave to Computer*, for example, includes more than a hundred diagrams – drawn by scientists and inventors such as Thomas Edison and Isaac Newton – that have substantially influenced the development of new ideas.

In recent years, diagrams have even found their way into children's materials. David McCauley's classic books, including *Castles* and *Cathedrals* bring life and analytical clarity to the structural relationships within buildings, and between the buildings and their inhabitants. The *Eye Witness* series provides children with visual representations of scientific and biological phenomena.

Richard Wurman has used diagrams, which he and other graphic designers describe as "information graphics," to illustrate a range of popular mainstream topics and places to visit (i.e., *San Francisco Access*). Edward Tufte has written a number of extremely influential books, including *The Visual Display of Quantitative Information*, which provide compelling uses and insights into the use of diagrams.

Newspapers and periodicals, traditionally text publications, now include diagrams on a regular basis (*USA Today* is a notable example).

Computer technologies, spurred by the realization that our modern culture is increasingly visually-oriented, have produced many programs that help users create and display diagrams, on topics ranging from personal accounting to public presentations.

We Are "Readers," Not "Writers"

Still, for most of us, diagrams are presentations made by other people. We may find them pleasing, but typically we don't even try to understand all of their dimensions. Sometimes this is because of a lack of interest. Often, however, it is because we really don't understand the general usefulness of diagrams, or we are confused by particular elements. Few of us have received any training in creating or interpreting diagrams.

Somehow our culture expects people to simply "pick up" reading diagrams; it believes that only particular individuals – such as graphic artists, architects, cartographers, or product designers – need to be skilled in designing effective diagrams.

Indeed there are, and always will be, particular individuals who excel in and professions that emphasize diagrammatic skills. Yet if diagramming is ever to become the pervasive visual language that it should be, it needs to be conceived as a way to exchange information, not just as a static display for others to read. That means that the basic conventions of diagraming need to be made explicit, and people need to have direct experience with these conventions. A language requires a fluid exchange of generation and reception.

66 I believe in giving the children experience with diagrams because then they have something they can use later for understanding more complex problems. I like to imagine that, years from now, they'll have all kinds of 'ah-hahs'..."
-Alison Quoyeser

Distinctions Among Diagrams

Diagrams, like drawings, cover a broad range of visual forms. Some are simply squiggles on a whiteboard. Others are carefully delineated computer-based renderings.

Some include a number of drawings, while others incorporate a great deal of text. Some are quite conventional – such as pie graphs or Venn diagrams or wiring diagrams – and others are thoroughly idiosyncratic.

Their unifying theme is that they are all visual forms designed to convey ideas. In this section we'll examine three common sets of distinctions that characterize diagrams: public vs. private, conventional vs. idiosyncratic, and presentational vs. conversational.

Public vs. Private Diagrams

Many of the diagrams that we see in the world around us are public diagrams. They have been crafted to convey a particular idea – to show how to put a toy together or to relay tomorrow's weather prediction, or to report on a political survey. In these cases, a substantial amount of time has been spent to create visual representations that are pleasing, informative, and accessible. Such diagrams must be understood by a wide range of people, so they need to portray data in a way that requires only a minimum of interpretation.

In contrast, many conference room whiteboards or personal sketchbooks contain diagrams that are more private. These have not been crafted to convey final results, or specific directions, or particular interpretations. They have been generated in a limited context – perhaps by one individual thinking alone, or by a small group working on a project together, or by a teacher talking to students. Private diagrams are designed to be "context-dependent"; that is, they are read and interpreted by the same people who create them.

It is similar to the contrast that exists between published textual works – novels, newspapers, movies – and personal documents – shopping lists, letters to friends, and personal notes.

Basketball Players

Notice the diagrams around you. Collect great diagrams, especially ones that you think explain big ideas simply, and spend time considering what makes these diagrams effective and appealing.

Conventional vs. Idiosyncratic Diagrams

There are a few basic diagrammatic forms that appear in most public diagrams. These include pie charts, graphs, tree diagrams, maps, charts, Venn diagrams, flowcharts, part-whole diagrams, and storyboards. Each of these forms has a particular set of rules for displaying information, although there is a great degree of variability possible (e.g., diagrams can be in color or black and white, they may include pictures or text or neither, they may be cartoonish or very serious looking, they may be ugly or beautiful). These conventional forms are used frequently in public diagramming sessions, since they provide instantly recognizable structures for coherently organizing information.

Many diagrams, however, do not rely upon familiar conventions. Instead, they are designed for a specific idea or situation. Though highly idiosyncratic, these diagrams can be brilliant at capturing a single idea in a very crisp way. Idiosyncratic diagrams are often drawn spontaneously to explain an idea, be it the landscaping of a garden, a movement in a ballet, or a new defense strategy on the football field.

	?	?	?
Idea	✔	✘	✘
Idea	✔	✔	✘
Idea	✘	✔	✘

I recently saw a diagram in a local newspaper that showed the differences between how the first half and second half of a championship basketball game were played. It depicted the origin of each shot that fell into the basket, labeled with the number of the player who scored the basket. This simple illustration showed clearly how one team's offense completely changed its strategy at the half; it also showed that almost all successful shots for both teams came from a very limited area on the court.

Because I had never seen a diagram like this before, it was quite idiosyncratic for me. However, this kind of diagram might well be familiar to basketball coaches. It might even be a conventional diagram, widely-used within this particular field. Many idiosyncratic diagrams become conventional as they are widely applied and appreciated.

Some teachers are extremely good at creating idiosyncratic diagrams, and over the years collect these to convey information to their students in diverse ways. Richard Feynman, for example, developed a set of idiosyncratic diagrams to explain his ideas about physics; Feynman diagrams are now central to the understanding of an entire generation of physicists!

❝ If you use a very formal (presentational) sort of diagramming technique early in the process...people will treat them as finished... You have to be careful to choose just the right level of finish, enough to get the idea across but not so much that it's misleading." -Scott Kim

The difference between conventional and idiosyncratic diagrams is tied to their arena of use. It is a distinction worth remembering as you learn to diagram, for it acknowledges that what you may take for granted in your own field won't automatically be recognizable to everyone. Sometimes you will need to invent something new to represent the idea you are trying to convey, and when you do, you should be explicit about your diagram's rules of presentation, so that others can understand what you're trying to communicate.

Presentational vs. Conversational Diagrams

Diagrams may also be presentational or conversational. Presentation diagrams are designed to match a general "publication" model, where their organization and conceptualizing have been done prior to publication. By the time they are made public – for instance, as part of a printed piece, or a speech, or a slide show – they are quite polished.

In contrast, diagrams used in a conversational context, with two participants or a small group, act as pictorial exchanges, typically in the context of idea development.

Often contained within an ARC cycle, conversational diagrams can be very iterative. The process starts when someone diagrams a basic idea. Verbal exchanges follow, and the diagram is modified, or another diagram is drawn beside it. More conversation follows. Other people may approach the diagram and add elements to it. More conversation follows. New ideas develop. Other diagrams happen.

This process continues until, at the end of the meeting, a final summary diagram has been achieved; it shows everyone what is agreed upon and what has been learned. As your ideas develop, your diagrams can become more detailed and refined. They can change from private to public diagrams. They can become slick presentational diagrams when you use them to explain your ideas to new audiences.

Certain kinds of diagrams are particularly well-suited for presentations. Others are excellent for eliciting conversation and review. On occasion, some diagrams are good for both situations. Thus you always need to be clear about the context before using a diagram.

It is also important to understand whether one is drawing a diagram to develop an idea, or to portray an established situation. Many of the best ideas are developed with fairly ugly diagrams that are incomprehensible. Many beautifully created and communicative diagrams are useful only within a narrow context and often curtail, instead of encourage, discussions.

66 A graphic conversation often takes place on little scraps of paper, but don't let that fool you. Pick up that napkin or whatever and paste it in your idea log, because it's a valuable record of how you do your thinking." -Scott Kim

The Process of Diagramming

You can use diagrams to help you think or solve problems, especially when you are working with other people.

Diagrams give you a place to put your ideas, to struggle with them as they develop. Diagramming can highlight a lack of preciseness in your thinking so that you go back and work out the details. It can help you avoid the unnecessary delay caused by waiting to do things until "you figure them out."

A diagram is not your thought, clearly. However, it is a concrete representation that you can point to as you develop your thought. Like an algebraic variable in an equation, it can help you to show what you do know about relationships, contexts, and meaning while you are struggling with what you don't know. It can let you get started even as you are not sure of where you will end. By showing the basic form of your idea, it can help explain that idea to yourself as well as to others.

Yet if you look at these markings later – for example, if you leave them for a while and then come back – you may not be able to understand them at all. More extremely, if someone not involved in your earlier discussion looks at your graphics, they will often be completely mystified. (Try walking through a school or university sometime and notice the writing/drawing on the blackboards. If you were not in the classes, most of these elements will seem like gibberish.)

The graphics you make are not the idea at all; they are just a mechanism to facilitate discussion and development.

The process of diagramming works as well for things you already know as it does for ideas in progress. In drawing a map to show a friend how to get to your house, you may discover that you don't know how far it is from the freeway, or what color that big house on the corner is, or the name of the street just before yours. By making the diagram you can find out what you don't know, so that you can pay attention and learn about it later. Many a teacher has learned new things by trying to explain what he or she already knew!

Diagrams: Managers of the Unknown

Diagrams can be extremely helpful when you are inventing something new, when there is no "reality" you can compare an idea with. If you're arranging your office furniture, for example, you can sketch a cluster of furniture near a certain wall before you know which specific pieces you will use. This is certainly easier than moving the furniture!

If you're planning your day, you can create a diagram that is empty in particular sections, highlighting all your free time. You can diagram the traffic patterns in your school hallways long before you specify the size and color of directional signs.

66 Mind mapping is useful for keeping track of ideas alone or in a group. Write your idea down and draw a box around it. As you get other ideas, let them branch out from this first idea, and enclose them in boxes or circles. Connect the boxes with lines or arrows." -Scott Kim

Diagrams as Placeholders for Ideas: Making the Uncertain and Incomplete Concrete

Diagrams can let you go a long way without knowing the specifics of something. In designing the *CD-ROM*, for example, we used a large number of visual placeholders to represent our design before we had the details worked out.

As an example, the entire DIAGRAMMING section was a set of cards composed of simple drawings and text, well into the last quarter of the development cycle. Even when the ENVIRONMENT, DRAWING, and IMAGINING sections were filled out, DIAGRAMMING was sketchy and vague. We built symbols for each of the sections, letting us encapsulate our ideas and view our entire product in context. After a while, we knew the subsections of DIAGRAMMING and the buttons for each subsection. As we learned more, we added it to the structure we'd established, a structure that was both a diagram on the wall and a prototype on the computer.

This worked quite well, although we ran into a problem toward the end of the design. At this point, we often confused our placeholders with the "real thing." One of us would criticize a subsection because the graphics weren't consistent or the narration wasn't right or the text was misleading. Then someone would remind us that these pieces were in progress and that modifications had already been decided, even if not yet implemented.

So be sure when you use diagrams as placeholders that you have a good system for discriminating these from real decisions, especially as a project reaches its final stages.

This Triangle and a Triangle: Objects and Symbols

In the visual arena, people will assume that a symbol is a picture, drawn to show a particular object in the world when it is really constructed to show a general concept. Often a diagram will stand in for something that is not yet known, yet we misunderstand the diagram and assume that it is a picture of something. You can see an example of this in a range of fields. It is particularly notable in mathematics learning, where general abstractions are at the heart of the inquiry, where variables are key to understanding ranges of similar situations.

Mathematics professors might draw a shape – a triangle, for example – to motivate a geometric proof in a class. Often this will be an isosceles triangle (two equal sides) or an equilateral triangle (three equal sides). The professors don't pay too much attention to these diagrammatic elements, because for them these triangles are symbols; they stand for "all" triangles.

Yet one will often find that students interpret almost all imagery as drawings instead of diagrams, believing the pictures are drawn to show something in the world rather than to explain an idea.

66 The graphic language that works best is simple, even whimsical. The goofier my drawing is, the more participation I get... I think *New Yorker* artists draw like second graders because they want that participation. They don't want realism; they want the lure of the partially filled-in framework."
-David Sibbet

Whereas the professor just chose a triangle that was easiest to draw, the students believed in all the specific attributes of the triangle as drawn.

This seemingly small difference in interpretation has a big impact on the mathematics learning that takes place. For one thing, a proof can be interpreted to apply only to a particular set of triangles. For another, the students completely miss the key notion of mathematical generality; they never realized that they were being given tools to use in understanding and predicting the behavior of all triangles, even those for which they had no specific dimensions.

Of course, how do you sketch "any triangle"? Any triangle that you sketch has particular dimensions and angles.

Welcome to the dilemmas of good diagramming! The convention for this kind of situation is to draw a triangle that has no particular regularities, but is just an ordinary everyday triangle (which, ironically, may actually be rarer in daily experience than "special" triangles like isosceles and equilateral). Alternately, you can sketch a number of different triangles, making it clear that you are discussing a class of visual objects and not any particular one.

Types vs. Tokens

In linguistics, this is part of a distinction known as "type" vs. "token." Consider the word *dog*. *Dog* can represent a type of object in the world (one that has four legs and barks and chases cats) or a particular entity in the world (Cheddar, my golden retriever/labrador mix that I take for a walk every morning). When I say "feed the dog," my children are very clear that I mean Cheddar, and not some neighborhood dog; yet if I say "look at the dog," while we're driving in the car, they know I am not referring to their pet.

Yet many people do not have enough sophistication in the visual domain to make this same kind of distinction when they are looking at an image. Visual images, particularly those that have any detail at all, are instinctively considered as referents to objects in the world. If they are sketchy, most people figure that the artist simply wasn't very good at making realistic drawings.

As we begin to use imagery as a tool for thinking and communication, we need to learn how to make these kinds of explicit distinctions in our drawings and diagrams. We need to show genetic "type" information in imagery just as we do with words. Until we do this, our visual language will be primitive and referential and it won't be up to the task of conveying important abstract ideas.

Diagrams and Drawing

Hopefully, you have a fairly good idea of how diagramming differs from drawing. Some distinctions are:

- Drawings are used to represent instances, diagrams are used to represent concepts.

- Drawings are often realistic, diagrams are editorial.

- Perception is critical for drawings, cognition is critical for diagrams.

There are some images that cross over between diagramming and drawing, that fall between the real and the imagined, the symbolic and the representational. Two of these are mentioned below.

Caricatures

Caricatures adorn posters, instructional materials, and the editorial pages of our newspapers. A caricature is a representation that derives its appeal from looking somewhat realistic while delivering an editorial point of view. It exaggerates the subject's physical features or personality traits (typically the negative ones) to emphasize this opinion.

Satirical caricatures of presidents and other politicians have provided many of us with amusement and insight over the years.

People are not the only topics that can be caricaturized. Saul Steinberg has created many wonderful geographic caricatures, including the "New Yorker's view of the world," where Manhattan and its landmarks are portrayed as large, primary elements, and all other states and countries are sized according to their relative importance to a New Yorker. Steinberg and others have carried this tradition to great lengths over a wide range of topics and forms.

The tension between realistic elements and editorial intent in these images is compelling, and an important technique for conveying ideas. Caricatures take advantage of the differences between drawings and diagrams, and straddle the line quite effectively.

Graphs

Graphs are widespread as visual representations go, though they are typically used in mathematical, scientific, and business domains. Most people don't understand graphs or find them boring and useless. Others use them constantly to explain their ideas to one another.

Graphs possess an interesting property that makes them useful as a form of visual communication: they can represent both general and specific information simultaneously. The points on the graph, at specific coordinates (such as [0,3] or [1,4]), illustrate particular observations or a point that satisfies a given relationship. The lines that connect these points show the overall pattern of these observations, the set of all instances that satisfy the relationship.

This duality makes these representations confusing to some and highly valued by others.

BIKE OR RICKSHAW

Using Diagrams to Spatialize Your Ideas

As you use diagrams, you'll probably find yourself pointing to the place on the board where you sketched your idea, sometimes in a form as simple as a dot or a squiggly line or a circle. "As I said over here," you may say, or "We have to remember this issue" as you point to the place where you jotted down a representation of the issue.

A diagram's ability to spatialize information can be a profound aid to your thinking. It lets you work out some parts of ideas without worrying that others are not yet done. It lets you "see" the connections among ideas and to explicitly show these relationships. It lets you group ideas according to various criteria, forming different categories on different locations on the board.

It takes things that occur over time and puts them into spatial arrays that lets one look at all the elements simultaneously in a structured view.

When you talk to people who diagram a lot, they'll tell you that diagramming and visualizing are both basically spatial abilities, and that their craft is mostly a matter of spatial arrangement, first in their mental images and then in their sketches. They'll say that coherent spatial organization encourages clear thinking, helping them to see what they know and don't know, and (literally) where the holes are on the page.

They will explain how it feels when an idea is actually taking shape on the board, when it comes "out of the air" and "onto the table" (or board) for elaboration and development.

Using Diagrams for Personal Note-Taking

You can use diagrams for visual note-taking – to highlight ideas, group issues, and emphasize relationships. Diagrams can reinforce your memory of a discussion by recording its structure as well as its content.

Usually when we take notes, we jot down a few words to capture an ongoing speech or conversation. This generally works quite well. As you bring diagramming techniques into your notes, however, you will find that you're automatically adding another dimension to the ideas, giving them structure as you are recording them. Putting this structure in will not only help you understand the ideas better initially, it will prove invaluable when you review these notes later.

Surrounding your textual notes with simple lines and arrows can turn them from a random laundry list into a structural document that offers an overall sense of the issues discussed and their interrelationships.

People who take notes frequently tend to develop sophisticated techniques. Some use different colored pens to record different kinds of ideas. Others leave a space in which to add comments later. Numbering systems are useful for showing the structure of arguments, and simple boxes or other shapes can represent relationships between ideas.

---

511 Watch Scott Diagram an Introduction to Diagramming
512 Consider Diagrammatic Dialogues
513 Listen to How Diagrams Are Useful to Groups
514 Distinguish Presentational Diagrams

66 In traditional note-taking, the content's there but it's not clear where to start... With visual note-taking, you can scan a page and immediately see how it's organized and where the important points are. It helps you pay attention not only to what the person says but the order in which they say it."
-Scott Kim

Talk and Diagramming

The diagramming process typically includes a great deal of talking. In fact, many diagrams are composed primarily of short phrases derived from a discussion, interconnected and spatialized in some way. You'll also notice that the verbal language people use while diagramming generally includes far more concrete references than other linguistic exchanges. "Let's look at this idea," "What were you saying over here?" "I don't understand this."

Before he or she invites comments and conversation, an individual diagramming an initial idea is almost putting on a performance. A verbal presentation that would be considered "a speech" without the diagram becomes something more akin to a mathematical proof where listeners are led down a garden path of logic by agreeing to certain assumptions, or a dramatic narrative, or a dance.

Scott Kim's introduction to diagramming in USING DIAGRAMS on the *CD-ROM* gives you a basic sense of the rhythm that is part of a diagrammatic presentation. As Scott tells you about diagramming, he also explicitly shows you what he means. These little diagrams serve as a record of what he has said, even as the words disappear into the air (though, in this instance, this is not a problem since you can repeat his presentation as many times as you want on the *CD-ROM*!).

You'll notice that there is an inherent spatial quality to Scott's "speech." As he says in his presentation:

"Think about not just what you write on the page – the words, symbols, and pictures – but also how they are orga-nized; whether they go top to bottom, left to right; how they're grouped, the different sorts of structures you might put them in."

The graphical elements are not simply tied to the verbalizations. Instead they form a structure in their own right, which together with the "talk," provides a unified representation of an idea, and creates a cadence and rhythm that is unusual for most verbal deliveries.

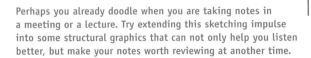
Perhaps you already doodle when you are taking notes in a meeting or a lecture. Try extending this sketching impulse into some structural graphics that can not only help you listen better, but make your notes worth reviewing at another time.

Diagrammatic Dialogue: Thinking and Exchanging

Diagrammatic dialogues usually work best with two to four people, all of whom are active participants in the development of the ideas and diagrams – whether they are talking, diagramming, or doing both. Group dynamics will evolve around the emerging diagram, often allowing a cohesion that is difficult to achieve when the energy is directed between people instead of toward a diagrammatic image.

You might begin a conversation with a few words and a diagram. Your colleagues could then respond with diagrams of their own or with modifications of your diagram. They might begin to draw on your diagram, adding features to make it better or pointing out critical points. If it is an effective conversation with the diagramming driving the ideas being discussed, the diagram itself will get more and more complex, and probably messier. Eventually, the diagram becomes joint property, and the ideas become part of the world rather than just part of your own musings.

The diagrams invite:

- additions
- elaborations
- alternatives
- and other interactions.

- subtractions
- criticism
- the noticing of new patterns

Often after a good diagrammatic dialogue, you will be able to construct a new diagram that represents your new understandings in a crisper, cleaner form.

Group Ownership of Ideas: Diagramming as "Group Think"

As the size of the group increases, the role of group dynamics becomes more critical.

Traditionally, diagramming plays a minor role, if any, in a large meeting. People talk about things, or look at prepared charts and lists. Political and bureaucratic issues can take over these meetings, since the ideas are verbal and can be ignored by those who don't want to hear them.

In a meeting that uses diagramming, there is less chance of this happening. A good diagram provides a group record of the meeting and the issues raised. Nothing gets left out or ignored. As a result, it can also reveal patterns and dynamics that were not consciously acknowledged before.

66 When we diagram a project as a group, it's like when the bomber pilots were going to take off, and they all set their watches. We're all setting our watches, we're all in the same time base, and we all know what the main goals are..."
- Frank Wiley 501 Overviews

This kind of "group graphics," as David Sibbet calls them, offers a basis for collaboration. It provides an explicit context for interaction between individuals and gives ownership of developing ideas to the entire group.

"In group graphics, the use of displays and diagrams supports the group process. The focus of group graphics is to mirror the conversation and thinking of the group. The discipline lies in learning what archetypal frameworks to use, the kinds of icons and ways of using words within those structures; it creates almost a kind of software tool for groups, operated in this case by a human being (a facilitator). Since computers have gotten so much more graphic, I sometimes introduce myself as a bionic cursor, and get a big laugh." David Sibbet

Effective group diagramming usually involves one or two facilitators. These people are often not concerned with the content under discussion; they may even be from outside the community that is meeting. Such consultants are used because they are expert at facilitating group discussions through diagramming techniques, and because they can maintain a certain objectivity about the meeting.

The facilitator's role at a meeting is to assure a thorough and honest record of a meeting, showing all the ideas presented. The final diagram, which is usually both attractive and well-structured, serves as a public record of the meeting's interactions.

Most of the time, group diagramming sessions can be structured to encourage coordinated graphic and verbal interactions. Just as some people speak often, some will draw often. A sincerely motivated group will recognize the diagrammers who can show everyone's ideas in collaborative fashion.

Formal Diagrams for Presentation

Some meetings, like those just discussed, are called to develop ideas, achieve collaborations, and arrive at group consensus.

Other meetings are more formal, such as when a small group summarizes its work for a larger group. Though these occasions may also allow a certain amount of feedback and collaboration, their format is typically informational, not participatory.

The diagrammatic expectation and standards for these kinds of meetings are very different. They require public representations that can stand on their own. They need to be meaningful outside of the context in which they were generated, and they need to be expressed through conventional displays so that outside viewers can read them.

---

66 To let a group collectively create meaning through the tools of drawing and visualizing is the most inspiring, activating, energizing kind of thing. It's become a passion for a large community of people that are now working globally in this manner." - David Sibbet

66 In meetings, each person usually spends time listening and trying to remember what everyone else was saying – or, more likely, what your own point of view is. If, instead, you diagram what everyone is saying, you create a sort of group memory …you can watch everybody's ideas develop." -Scott Kim

Some Diagramming Techniques

Whatever your interest or level of ability, whether you would like to become better at having graphical dialogues or just more effective at taking personal notes, there are certain basic techniques that can help you.

A number of these technique can be found in exercises on the *CD-ROM*. These are based on a consideration of diagrams in terms of three primary elements.

One of these is drawings, pictures that show specific objects or object classes found in the world.

Another is words. Diagramming often includes key words or prominent phrases to represent developing ideas, or textual references to other sources, or verbal labels to make a drawing less ambiguous.

A third element is symbols, images that have been designed deliberately to represent an idea in an abstract manner.

Visual Symbols

Many symbols are used spontaneously in diagramming. As you are describing your day, you might draw a line to show the whole day and then draw squares to show just the main events. As you generate this diagram you might say, "So if this line is the whole day, then this box shows when I went to the store; this one shows when the accident occurred; and this one is when I got out of jail."

521 Introduction to Symbols
522 Identify the Meaning of Different Symbols
523 Communicate an Idea with a Symbol

66 A symbol is a simple picture that stands for an object or an idea. Symbols can be more or less abstract...they can be public or private... Some public symbols are so widely accepted that they've become a sort of universal language." 521

You can create these symbols spontaneously, then use them to stand for your ideas during the rest of the conversation.

This symbol-making activity is an activity used frequently by scientists and children, but not as readily by the rest of us. Scientists will often assign a squiggle to an idea as they work it out; they may also assign it some kind of mathematical variable or other conventional form. They will begin with phrases like "If we suppose that" or "Let x stand for," moving quickly into hypothesis. Children, particularly in early play-acting, will tell you a story in terms like "the chair is the kitchen and the teddy bear is your pet; now you should feed your pet." In many cases they seem so entranced with their symbology that it becomes quite real for them (and for you if you are lucky enough to share their imaginative ramblings).

As we get older we seem to think we should know more things, including conventional symbols, and so we do not make symbols up spontaneously.

Yet this is a critical activity within diagramming. For you will frequently be challenged with the task of graphically showing an idea symbolically within your diagram. Short textual phrases can serve you as well in an extremely large number of conversations, as can dots, squares, circles, and squiggles.

Yet as you gain experience in diagramming, you will find that the effectiveness of your conversations often depends upon the quality of your diagrams, including the clarity and meaningfulness of the symbols you use. This is particularly true when you try to engage newcomers in your diagrammatic discussions or when you present your ideas to a larger audience. So you need to learn to make effective symbols if your ideas are to move beyond a single simple conversation.

66 Private symbols are often meaningful only to the person who drew them. Most symbols require some amount of context – from a simple label to a verbal explanation – to make their meanings clear." 521

Some Diagramming Techniques 149

International Symbols, Computer Icons, and Hobo Signs

All of us encounter graphic symbols in our daily lives. Traffic signs, no smoking signs, and restroom signs, for example, all employ symbols that convey information without using a specific verbal language. Many are part of a larger set of "international" symbols, conventional symbols that are used and recognized around the world. These fall into the category of public symbols because they can stand on their own with little or no interpretation.

Many people were introduced to the serious use of symbols through the Macintosh computer. This widely-used tool created a visual (instead of textual) environment, establishing a set of "icons" to help users navigate through, organize, and access information. Many of these Macintosh icons are public – the trash can, for example, is easily recognized by anyone, as a symbol for disposal. These commonly known icons are simply objects in the world that have been given symbolic status in the computer domain.

Other graphical symbols are far narrower in context. Many of these have developed within particular domains, such as agriculture, geology, meteorology, architecture, and engineering. Easily recognized by people within those disciplines, they are mystifying to outsiders. Some of these symbols have varying meanings, such as a circle with a dot in its center, which represents "the sun" in astronomy, "adult" in biology, and "gold" in alchemy.

Certain symbols are particular to an even smaller, cultural niche. Among these are "hobo signs"– a set of secret symbols developed by traveling hobos, used to leave messages of encouragement or warning to other hobos who pass through the same geographical area.

In order for a public symbol to be successful, its definition has to be agreed upon by a large number of people. Once this happens, it can become an effective convention. This can be especially complicated if the symbol represents not only an object or an idea but also an action. For instance, some people don't understand the trash can icon on the Macintosh. They recognize it as a trash can, but they have no idea that you can drag items to this icon and then eliminate them. What's also interesting to note, however, is that once people know how to use this item, they have a difficult time understanding why others don't find it obvious.

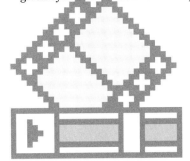

Which of the symbols in Symbols Match do you find best at representing their intended meaning? Make a list of the top five in your Sketchbook. Also make a list of the worst five. Compare your list with someone else's to see how your lists vary, and how your reasoning differs.

This example can serve as a useful reminder that one person's symbol is another's meaningless squiggle. Communities of use must be built around public symbols in order for them to be effective outside of their context of origin.

Private symbols, on the other hand, have only to satisfy the needs of the person creating them and their intended audience.

Effective Symbols

Making up your own symbols can be a lot of fun. Whether you generate them spontaneously or after hours or days of thought, you will probably be pleased with the direct manner in which you've represented complex ideas. Careful, however: you may be chagrined to find that to others your visual symbols are confusing rather than clarifying, disorganized rather than pleasing. Unless your diagrams are for purely private use, you will want to consider others' ease of interpretation as you create your symbols.

To get a sense of this, the SYMBOLS MATCH exercise on the *CD-ROM* includes a wide range of frequently used symbols that are presented to you out of context. Looking at these symbols and trying to identify them should give you an understanding of what it's like to encounter other people's symbols, and how ambiguous they can be when viewed for the first time.

As you try this exercise, notice when and why you become confused. What graphic forms work best for you? Before you look at the possible definitions, try guessing what the symbols stand for; place your hand over the choices on the screen to ensure that you don't peek at the answers. Trying this with a friend or colleague can also spark a discussion about what makes a good symbol.

Reflections

How many symbols in this activity were familiar to you? How did your interpretation of these differ from those you had never seen before?

What did you think of the "distractors," the incorrect answers included for each symbol? Did they make it harder or easier to identify the symbols?

Play the game a second time. Did you learn the symbols in one try? Were some harder or easier to remember? Why?

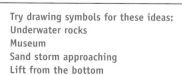

Try drawing symbols for these ideas:
Underwater rocks
Museum
Sand storm approaching
Lift from the bottom

Creating Symbols

Now that you have criticized and admired other people's symbols, it's time to create your own.

In the SYMBOLS DRAW exercise, you have a chance to work with a friend to tune up your graphic communication skills. Much like a number of popular visual games, this activity asks you to graphically depict a phrase so that your friend can recognize it. This game provides a set of alternatives from which your friend can choose; this makes it a bit easier to play.

The surrounding context for a symbol greatly influences the form you need to give it. Creating a symbol that is to stand alone is a much different and more difficult task from one that can be identified from among a few choices.

Spend a lot of time on this game, honing your skills. Keep score if this engages you, by recording how many tries it takes each of you to guess correctly – or try giving yourselves a time limit by using a stop watch or timer.

Your goal is not to confuse an opponent by drawing terrible symbols, but to help your friend succeed by drawing as clearly as you can. It is a collaborative, not a competitive activity.

If you finish all the lists provided, make up your own. Or you might simply use the drawing tool to spontaneously sketch an idea. Always ask a friend to interpret it, since you want to tune your skills in communication, not just in drawing attractive sketches.

Reflections

What did you learn about communicating with visual symbols? What techniques did you pick up?

You have probably found that you can make almost any symbol work for you in context. You can draw a completely ambiguous shape and simply assign it a meaning and be done with it.

The implication here for your diagramming is that, in cases where there is little context, you need to make more communicative symbols. Also, as the number of symbols you use in a diagram increases, you will want to become better at making distinctive symbols.

Keep practicing and testing your intuitions about symbol-making with others. Unless you don't expect anyone but yourself to see them, don't fool yourself with symbols that you create in isolation.

---

66 If you want to learn how to do idea diagramming, you need to think about not just what you write on the page – words and symbols and pictures – but also how they're organized… the different sorts of structures you might put them in."
-Scott Kim

Diagramming Structures: Showing Relationships

The foundation of a diagram is the structure into which its elements – the text, the drawings, and the symbols – fit. For it is the structure that shows the relationships among these elements.

A well-constructed diagram displays patterns, high-lights similarities and differences, and allows the spatialization of ideas in ways that enables these ideas to be sorted and summarized.

Provocatively, there seems to be a small number of basic organizers of information. Richard Wurner suggests five basic organizational structures: Linear, Alphabetical, Time, Categories, and Hierarchies (LATCH). It is impressive how many things can be organized in these ways.

The formal structures underlying these – each of which can have many surface forms – include the following, which appear on the *CD-ROM*:

- Linear Structures
- Clusters
- Hierarchical Structures
- Matrices

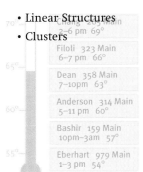

Linear

Linear structures are used to organize elements that vary on one dimension. A timeline, for example, is a linear structure that shows the occurrence of events over time. The location of a house on a street can be shown in a linear diagram, with the street represented by a line and the houses depicted as locations along this line.

Lists of all sorts are linear. Sometimes the relationship among items on a list is ordered according to a single attribute, for example a list naming the world's mountains from highest to lowest. Other lists are composed merely of items that come one after another; for example, the next part required to make a particular piece of machinery, or the next item on your grocery list.

Clusters

Cluster structures put related items into groups or categories according to a specific criterion, which may be named or remain unnamed. This relationship is the basic "set" structure of mathematics, which describes how certain items belong to a set, while others do not. Clusters can also include other clusters or overlap other clusters.

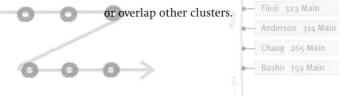

> **66** Linear structures, like lists, arrange items in lines. Sometimes only the order matters. Sometimes the exact position along the line also matters. The line may be vertical, horizontal, curved, or interrupted. Labeling the line can help how items are arranged." 532

An example of a basic cluster diagram would show two groups – dogs and horses. One cluster would contain the different breeds of dogs, the other different breeds of horses. Within the dog cluster the breeds might be further grouped – with one group showing, for example, only herding breeds or terriers. Horses might be separated into hunters or racers. Both dog and horse clusters could also fit into a larger cluster titled "four-legged mammals."

Clusters can be shown graphically in many different ways: they can be shown in different circles; or they can be grouped closely together; or they can be united by the same color.

A well-known cluster structure is the Venn diagram, a diagram that is frequently used to show two groups of elements with overlapping membership. Tall trees are in the overlapping area of a Venn diagram that shows tall things in one circle and trees in another overlapping circle, for example. These double overlapping cluster diagrams are very helpful in showing a wide range of relationships. They also give people a tool with which to derive unique meanings. For example, in Alison Quoyeser's second grade class there is often a big Venn diagram on an easel waiting for the children when they walk in. Each of the two circles would be labeled for example, "I brushed my teeth this morning" and "I have one sister." As the students enter, they sign their name in the correct region. This is not only fun, but it also teaches students to use Venn diagrams and to interpret the data within them.

Hierarchies

Hierarchical structures let you show sets that belong to other sets that belong to other sets, and so on. They are actually a special subset of cluster diagrams.

Typically arranged in a "tree" format, these diagrams are helpful in showing such relationships as corporate hierarchies and family structures. They can also be very informative in showing part-whole relationships of machines, organizations, or concepts. Category-example relationships also can be shown directly in tree diagrams. Alternate formats for this structure include a main idea-subsection outline and nested clusters.

66 ... You can cluster items by moving them near each other, by aligning them, by drawing boundaries, or by marking them with distinctive colors or other visual cues..." 533

Matrix

Often items are organized according to more than one variable. A person may be female and have red hair, for example, or a project may be highly risky and have large potential for profit. A matrix helps you organize elements according to two or more linear variables.

Typically, two-dimensional matrices are shown in a set of columns and rows, and are often called "tables" or "charts." A well-known two-dimensional matrix is a Cartesian graph, which shows the values of an element on two-dimensions: the x- and y-coordinates. This format provides an alternative to the table for displaying such data and is most useful when you have a large number of points and are seeking a general pattern.

A Structural Overview

None of these four different structural forms is particularly complex, yet they can be used to give coherence to a wide range of ideas. In fact, just discovering which structure best organizes your ideas can sometimes lead to major breakthroughs in your thinking.

Look for these forms around you, in public signage, in print pieces, and in people's personal notes and diagrams.

Once you are aware of these structures, you'll probably find them everywhere, in a wide range of styles and a range of problem areas that vary from the profound to the trivial. You'll find that they offer a general structural vocabulary with which to organize your thoughts and to arrange the basic elements in your diagrams.

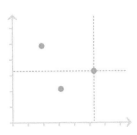

66 Hierarchical structures put items on the branches of a tree... You can draw hierarchies growing upward from the bottom, downward from the top, from left to right, or radiating out from the center..." 534

66 Matrix structures, like tables, put items in rows and columns... Tables are useful when you want to make a list of items, each of which has several attributes... Matrices are useful...when you want to combine lists in all possible ways." 536

Diagrammatic Composition

Conversational diagrams are usually composed of a number of elements and structures. A diagrammatic dialogue may include a range of scattered text phrases, drawings, and symbols, and more than one conventional structure.

The task in making a diagrammatic record that can support the ongoing discussion is to compose these elements to show coherent relationships. This coherency will ultimately be useful in enhancing participants' memories and in conveying the results of a conversation to others.

A range of compositional techniques exists to assist you in producing coherent diagrams.

The *CD-ROM* presents three:

- Position or layout
- Grouping
- Emphasis

Position or Layout

The basic location of items in a diagram provides the first level of organization. Those elements that are close to each other are probably related to one another, and those that are farther apart are probably not as closely related. The placement of elements on top of each other (or overlapping or side by side) conveys a particular perspective on the relationships of those elements.

Often such organization happens naturally during a discussion. At other times, you'll find that you need to return to a diagram already created during a discussion to redraw it, incorporating spatial organization that makes the relationships explicit.

The POSITION exercise asks you to convey different ideas by positioning basic shapes on the computer screen. The challenge lies in being able to only manipulate spatial position. You aren't given other attributes to change or manipulate as you would in a normal diagramming exercise.

You may find this task hard initially, noticing how much you rely on text phrases to explain yourself. You may be tempted to try and "draw" images with the squares, instead of expressing an abstract concept. But with practice you'll discover how much you can convey through simple arrangements.

You can extend this exercise by making up your own examples, and arranging these on the screen, or with paper and pencil, or with cutout squares. You might also try this exercise with more or with fewer elements, or by using more than one shape.

Emphasis

You need a way to show which elements you consider to be important in a diagram.

One way to emphasize something is to separate it from other elements. Another is to vary the size or line weight of an element. The larger you draw or write something, the more attention you bring to it; similarly, the thicker you make the letters or the outline of a shape, the more emotional weight you give them.

You can try this in the EMPHASIS exercise on the CD-ROM. This exercise allows you to change the sizes and line weights of basic shapes. Organize these shapes to convey a basic idea to a friend. Change the size of the shapes and their line weight to emphasize certain aspects and lend additional clarity or complexity to your communication.

Make up your own phrases or concepts and depict them on the screen. Explore your own ingenuity and see what you can do with these primitive elements.

Grouping

Another basic technique you can use to convey an idea is grouping. You can show relationships among elements by putting them in the same group or by making connections between them.

In the GROUPING exercise, the two major elements remain stationary. You express their relationship to each other through a range of graphic indicators – including boxes, circles, arrows, and connecting lines.

Reflections

Although the compositional elements you've been playing with are extremely limited, you can use them to communicate a wide range of ideas. Think about how you might combine all of these elements – position, size, line weight, and grouping – to convey your ideas. Design a few two-person exercises for yourself on a topic that interests you, using all or some of these elements to represent your ideas.

Begin to notice how public diagrams use some of these basic elements. What other elements, structures, and techniques do they rely upon to convey information? As was true of drawing, your job is to notice and then to incorporate the different visual solutions you notice into your own diagrams.

66 When you diagram ideas, it's important that you pay attention not only to what you draw, but also to how you compose it on the page. You can...communicate different messages by varying visual characteristics such as size, line weight, color, texture, or orientation." 541

Putting It All Together: Show and Tell

It's time to combine all these elements and techniques for communicating in diagrammatic form.

The SHOW AND TELL exercise on the *CD-ROM* offers you a variety of tasks that require you to create diagrams using all the techniques you have learned.

Maps: Showing Geographic Spaces

A map is an interesting type of diagram. Formal maps, which need to portray geographical information, and to work for all people and all situations, follow cartographic standards and guidelines. They include explicit conventions for rendering topography, roads, and landmarks.

When you make a map for a friend, however, most of these formal rules do not apply. Your task is to create a diagram that will get the person where he or she wants to go.

The design of this informal, customized diagram is based on what your friend knows and doesn't know. If she knows the freeway exit, you have a simpler task than if you need to show her how to find the freeway itself. If he already knows the location of a nearby shopping center, you have a different starting point than if he has never been to the area at all.

Your map needs to include the detail that is required for navigation and leave out everything else. For example, you might highlight one or two stoplights that appear before a decision point, but certainly you don't need to show all the crossroads along a thoroughfare. You need not say exactly how many miles must be

Draw a map of how you reached your last vacation destination.

travelled, but you do need to estimate the distance (perhaps in minutes) before they reach the next key decision point. Absolute scale is not necessary, but relative scale –"it's a long way along this road"– should be indicated.

You'll probably want to include some perceptual information, such as landmarks: "Turn left at the big white church," or "Turn right just past the red carousel," or "Keep an eye on the town hall bell tower and aim towards that." The diagram would need to include the church, the red carousel, or the bell tower. You might use size to emphasize particularly important landmarks or decision points, or you might use arrows to point these out. You might want to generate a list of directions to accompany the map, or you might want to group the trip into three main sections (getting to the freeway, getting to the neighborhood, getting to my house) for clarity.

Try some of the map-making tasks on the *CD-ROM*, and look at the examples provided to get a sense of the elements and techniques you can include in your own diagram.

Events

Diagrams are also very useful for showing events. A basic timeline can be invaluable in revealing what happened during a particular period, and when. For example, Columbus came to America before Gutenberg invented the printing press, but after Joan of Arc died.

A timeline can also show just what is planned for a particular event; the bride will come to the church at 10:00 A.M., will walk down the aisle at 10:30, and begin to live "happily ever after" at 11:30.

A timeline can be embellished extensively with drawings, symbols, and other structures. Objective elements can be included, as can subjective impressions or emotional reactions.

Diagrams can also show sequences of quite complex events, like what goes into hitting a tennis ball, or how you got into your current profession, or the steps you have to take to get a driver's license.

Diagram your failures and successes in the last year.

Diagram your school career.

You might want to start by simply diagramming sample days in your life, to get a sense of how to show events. You will probably notice some patterns emerging that you were not even aware of. This ability to make hidden patterns and concepts more apparent is one of the great benefits of diagramming. We've included some examples of other peoples' diagrams on the *CD-ROM*.

Life: Showing Personal Views

Diagramming events in your daily life can be a unique resource whose value extends well beyond merely helping you to note the order of events during a day. You can draw a diagram of your current job, your closest set of friends, your childhood family, or your plans for the next 10 years. Each of these diagrams offers you the opportunity to focus and reflect on different aspects of your life. You may gain insight just by noticing how you diagram these personal perspectives.

You might notice, for example, that your ten-year plan shows no activity in the three-to-six-year period, because you expect current activities to produce great changes only in the seven-to-ten-year period. Or you might find that you draw one of your sisters close to you and the other farther away, or that your mother is large in your diagram and your father relatively small.

There are a number of "art therapists" who specialize in helping people understand themselves through diagrams such as these. However, you can do a good deal of this for yourself, moving around and studying the elements in your diagrams to gain a sense of what is troublesome to you, or using your diagrams to confer with friends about things that are important to you.

Try representing some aspects of your everyday life, and spend some time reflecting on what you have diagrammed. Also look at the examples on the *CD-ROM* provided by other people who attempted some of these same diagramming tasks.

Ideas: Showing Relationships and Concepts

As has already been stated many times, diagrams are wonderful for representing ideas. Now is the time to practice your techniques, combining all the different perspectives you have experienced in this chapter. Try the tasks on the *CD-ROM* and notice how others have represented these concepts.

THE MORE YOU EAT THE BIGGER YOU GET

Diagram your day.

Reflections

In working through these exercises, you are gaining experience in the use of basic diagrams for communication. In many of these activities, however, you have been operating in isolated situations. Be sure that you also practice creating diagrams in natural contexts that involve colleagues and friends who can give you feedback. And remember that diagramming will be especially effective if you depict topics that are important to you.

As you do this, pay particular attention to developing fluidity and expanding your repertoire of techniques. Become comfortable with this process. Be playful as well as reflective, attempting both simple and complex topics.

Dynamic Diagrams

As you become increasingly skilled in diagramming, begin paying attention to the dynamic aspects of your diagrams. Notice particularly your "performance skills" – such as the rhythm you apply to your diagramming process – alone and with friends.

Notice how your diagrams develop over time. Watch as they move through concise and messy phases, literally, in their visual form and in the underlying concepts they represent. Note how some diagrams are incomprehensible until they are completed, while others begin quite crisply and develop in a cohesive manner.

Once you have developed the basic technical skills, you might also want to begin experimenting with diagrams that exist over time – that is, animated diagrams. Illustrators and movie producers have created wonderful sequences of diagrammatic elements to show ideas. Even very simple sequences, such as those shown in the STRUCTURES section on the *CD-ROM* can convey information in ways that static images cannot. Simple computer animation programs can help you in creating such representations.

You might also want to practice diagramming sequences without the use of animation. Simple cartoon or storyboard conventions offer many ways to represent ideas and events that are supposed to be taking place over time.

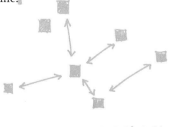

MANY PREVIOUS JOBS

Make a diagram of your family.

Diagram this adage: The more you eat the bigger you get.

Because they are very handy and convenient, whiteboards, pieces of paper, pencils, and pens are now typically the tools used to create diagrammatic conversations. As technologies develop, it should be just as easy to diagram dynamically. If you have mastered the basic techniques of thinking in diagrams and of representing your ideas in these forms, you will be ready!

Interface Design: Diagrams Over Time and Space
Although technologies that support facile diagrammatic conversations are still under development, personal computers have demonstrated that diagrams do indeed already have a place in the electronic arena.

The Macintosh desktop interface is itself a grand experiment in diagramming. It is based on a few conventional structural elements and a range of carefully designed symbols. It extends the use of dynamic diagrams very deliberately into the interactive domain.

The development of multimedia programs has furthered notions of "spatial interface;" the notions that (1) a diagram can act as the basic organizing structure for a program, and that (2) content materials can be accessed in non-linear sequences.

Indeed the design of the entire *CD-ROM* has been one big diagrammatic exercise. Debates about which elements to include, how to articulate them, and where to place them abounded throughout the whole process. Now, as a user, you are accessing this diagram in a variety of directions, interpreting it in your own way.

Ancient peoples drew diagrams in the sand and on cave walls. During the past few centuries, we seem to have forgotten their significance as we have been entranced by the genius of the printed page. Yet, our newly emerging technologies are making it possible for us to again rely on visual representations and diagrammatic organizations to express ourselves and to organize our thinking.

When I was a child, I diagrammed a lot. I thought of it as my secret weapon for doing well in school. I sometimes wondered if it was fair, but I always made diagrams of my math problems and science projects, and it helped me succeed. I learned to use diagrams with others only when I became a postdoctoral fellow in architecture. Then I delighted in the fun that I could have with others creating brand new ideas in diagrams on blackboards and restaurant napkins.

During the past ten years I have learned that there is an entire community of visualizers who diagram, in corporate meetings as well as engineering reviews, in sales presentations as well as event planning sessions. When permission is given and a milieu for interaction is established, diagramming can provide a rich venue for communication and invention.

And it will only get better as newer technologies become available, and as there are more participants in diagramming activities.

Yet, all this facility in seeing, drawing, or diagramming is not really useful unless someone has something imaginative to say.

Let's roll the *VizAbility* cube one final time as we consider the process of Imagining.

Imagining

Learn new perceptual abilities and imagine how things might appear. Envision objects in two and three dimensions. Imagine transformations of visual forms.

Gain experience in interpreting and using ambiguity in productive ways.

The world lives in our mind's eye. We can consciously enhance our thinking and inventiveness by engaging our imaginations.

Learn basic brainstorming methods to increase the number of ideas you can generate.

Learn to take advantage of what is silly to discover what is profound... **601 Introduction**

Introduction

My daughter saw a sign in a computer store the other day. On it was the following quote from Albert Einstein: "Imagination is more important than knowledge."

She asked me why, if this were true, did we all have to go to school? As is often the case with children, her question was the right one; I had no quick answer – only a couple of reflections.

First, I was pleased that she had seen this quote, and that she realized someone as revered as Einstein valued imagination.

Second, I wondered why she thought of school as a place of knowledge rather than of imagination?

I mumbled something about how school really is a place to use imagination and that she should keep using hers, and we went on with our day.

But I was unconvinced by my own answer. Why, I asked myself, do we think of schools and libraries as places "to hand down the knowledge of the world"? Why don't we think of imagination as something of equal value, that also needs training and cultivation? As I considered all this, it seemed quite odd to me that I had never received any training in using my imagination, even though my imagination is certainly my most valued and productive ability. Without it I could never have even started a product like this, or qualified for a job at Apple Computer, Inc., or have done well in school. And it seemed quite odd to me that our culture doesn't encourage this skill explicitly, in our schools and in other venues.

611 Transformations
621 Block Builder
631 Guided Fantasy
641 Brainstorming
651 Magic Theater

66 It's my belief that a lot of science gets developed by using your imagination, your internal imagery, and manipulating it. Einstein talked about the development of both special and general relativity. In both cases he was asking a very simple visual question... They are simply imaginings about the world."
-Rob Semper

The Nature of Imagination

It became clear to me that as we go about our every-day lives, we often forget to spend time enhancing our imaginations. We think of imaginative people in a separate category from the rest of us, somewhere out there with the "creative" and "artistic" and "talented."

But imagination is an integral part of all of us from the beginning. If you've spent any time with children, you already know this. Children are brimming with imagination, curiosity, and playfulness. Yet as we get older, most of us shy away from this part of ourselves, moving toward more serious and concrete issues. We dismiss imagination as something fanciful, illusory, even delusional.

For all that, we still use our imaginations every day. We plan what we are going to do, and we think about what might happen tomorrow. We analyze what steps to take by considering a range of possibilities. We imagine what it would be like to drive that flashy sports car, hit that home run, have that person as a friend, climb that mountain, or finish this chapter.

Imagination is one of the most human of our capabilities, one that we automatically point to when we distinguish ourselves from other animals. It is among our "higher order" thinking skills, the ones that enable us to invent tools, to solve problems, to innovate, to look ahead to the future, and to hope.

An Attitude

Being imaginative is an attitude. It doesn't typically take special equipment or expensive consultants or financial grants. You might not be able to deliver on your imaginings without other resources, but it is typically not very expensive to come up with ideas, and to define something that is worth pursuing.

Perhaps it's because it *is* free, and because even young children have it, that we don't value or cultivate our imaginations more. Which is too bad, because if we did, there are huge advantages to be gained.

On a profound level, we might come upon a new paradigm, shifting scientific understanding or human philosophy. We might build the next billion dollar toy, or create a masterpiece of art, or choreograph a brilliant new play. On a more mundane level, we might save our business from financial disaster, or design a better house, or find clever ways to resolve an issue with a client. In a personal context, we might make a new friend, produce a great dinner party, initiate an exciting conversation, or plan a wonderful vacation.

Many of the things we acclaim, respect, or financially reward in our culture were produced and maintained by people with great imaginations. But until they receive those accolades, the source of their accomplishments – their imaginations – often goes uncelebrated.

66 I think imagining is seeing through the darkness, seeing through the walls, through the mirror. Imagination is being able to, after you've seen through it, to step through it."
-Frank Wiley 601 Overviews

66 I think everybody has an imagination but it may be that in their lives it hasn't been valued and exercised, which is really sad. If children know that you value their imaginations, they will shine. They will emerge as really strong individuals, certain that they have something important to give to the world."
-Alison Quoyeser

Tips of Icebergs and Other Distracting Metaphors

There are a number of reasons for under-valuing imagination. For one, it has a reputation of being highly impractical. Imagination may result in some delightful daydreams but it doesn't get the spelling homework done. A new invention may be highly imaginative, but it's not affordable or maintainable, and it's hard to get parts.

Imagination also has a reputation for making a great start but a poor finish. Team projects may display a great deal of imagination at the early stages, but pragmatics soon take over. The dreamers step aside and the do-ers step in to put in the real work. After all, we've been told repeatedly that success is 1% inspiration and 99% perspiration.

In addition, imagination typically does not bring the same financial reward as implementation. Compared to the cost of the materials and labor to build a project, architects receive a relatively small fee for their design. Research groups in corporations are funded at a much lower level than development groups. And, in many industries, product development is funded at a much lower rate than advertising, marketing, or sales.

There is a certain inherent logic in all of these situations. Materially, it doesn't cost much to sit and think, or to make some drawings and diagrams to explain a new idea. Indeed, by most measurements, imaginative invention is only the tip of the iceberg in an overall activity. The remaining problems, costs, and hard work lie waiting beneath the surface, vast and expensive.

There is some short-sightedness to this approach, however.

Most obviously, there isn't going to be much development work unless there is a good idea to base it on. There aren't going to be production problems unless there is an initial design worth pursuing. All the money in the world isn't going to turn out a great product if the idea at its core is mediocre.

Imaginative activity doesn't necessarily only take place up front, either. Successfully distributing a product requires a good deal of imagination, as does coordinating materials for a building, or figuring out a new piece of packaging.

66 You can't depend on your judgment when your imagination is out of focus." -Mark Twain

If we really think about it, we can see how our usual distinctions – between people who are imaginative and those who are not, between stages of a project that require imagination and those that don't – are ultimately self-destructive. We can all benefit greatly from a culture where everyone is imaginative, where everyone thinks of new ways to do things, where everyone incorporates dreams about how things can become better into their daily lives.

And so it is time for us to place explicit attention on our developing imaginations, to take the time to understand them, to enrich them, and to practice using them. For as you add a fresh imagination to your other visual abilities you can bring a powerful set of perspectives and capabilities to your world.

The Image in Imagination

As the popular expression, "in the mind's eye" reveals, we often link imagination to the general idea of "a mental image." Our concept of imagination is closely tied to that of perception. At one extreme, the construct of "a mental image" may actually imitate perception: it invites us to see an object even when the object isn't there.

More typically, while people describe a similarity between their mental images and their perception of real objects – ascribing perceptual features and a sense of immediacy to both – they also say that imagined images differ by being more sketchy, vague, and abstract.

Extended arguments about internal images and their comparison to perceptions abound. Psychologists question the fundamental nature of mental imagery: they wonder about the nature of its functioning, its physiological basis, its usefulness in thinking, and whether it really exists.

As these weighty arguments rage, people continue to form mental images, by accident, for amusement, and to enhance their professional work.

Try it yourself. Take your time with each of the following items. If it helps, close your eyes after you read each phrase.

Imagine:
- a clean magenta cashew
- a fragrant ivory couch
- a smooth crimson baby
- a warm styrofoam car

703

66 We are all of us imaginative in some form or other, for images are the brood of desire." - George Elliot

Now try making mental images of these more personal experiences.

Imagine:
- your mother's face
- the dashboard of your car
- the front of your house or apartment
- yourself brushing your teeth this morning

Let's make this a bit more dynamic.

Imagine:
- a horse running across a field, kicking his heels in the air
- a trapeze artist flying across the sky
- Willie Mays catching a fly in center field
- a space shuttle taking off on a cool morning

Now let's get a bit more whimsical and flexible.

Imagine:
- being born old and growing younger
- wearing your shoes as a hat
- sitting on a ball and bouncing a chair
- swimming in the air

703

What do you think? Is your experience of a mental image – whether it be an object or an activity – the same as perceiving it directly?

Most likely, you will find that creating these images is quite different from actually seeing these things. The constructed images may seem more abstract, possessing a general quality that makes them more like "dreams" than "photographs." Some individuals, however, will notice a similarity between the images of objects and scenes experienced and those constructed purely for these exercises. Both sets may contain so much detail, and they may seem so "picture-like" and realistic that these people may question whether the images aren't exactly the same.

An Imaginary Horse

As a further example, imagine a horse standing in a field. Now imagine the horse walking toward you. Notice the grass blowing in the wind, and the sky.

Got it?

Okay, now think about your image. What color is the horse? Brown? Black? White? Dappled grey? Is it a pinto?

Was it any color at all to begin with? Maybe not. Did it assume a color as you thought about it? It's likely that as soon as you wondered what color it was, the horse became a particular color. Or maybe it didn't, and you wondered for a moment how there could be a horse, even one in your imagination, that wasn't colored.

Then again, maybe you saw the horse and its color clearly all along. Maybe the horse turned brown when I asked you if it was brown, and then black or white or dappled grey when I asked about these colors. (Of course, if you don't know what a dappled grey or a pinto is, your image may not have changed much with these questions.)

The imaginary horse can be any color you want it to be, when you want it to be. There are no right or wrong answers. You have great personal freedom to manipulate your own images. You can do this in a frivolous way, just for fun and fancy. You can also use this skill in serious activities, where imaginative flexibility is an advantage.

Now think of your horse again. Did it run toward you or walk? Can it do both? Did it come from the left or right?

Was there a fence around the field? What did it look like?

Was the grass lush and green, or perhaps golden (and tall, like wheat)? Was it short or long? Or was it unspecified?

What color was the sky? The deep blue of autumn or the hazy blue of summer? Did it indicate a particular time of day, such as sunset or noon? Were there clouds? What kind?

What other things might you imagine in this scene? Maybe a small foal running beside your horse, perhaps on the left side. Maybe spring flowers in the meadow and rolling foothills in the distance. Or, in complete contrast, maybe a meadow full of old abandoned cars. Or one crisscrossed with gullies carved by erosion.

The whole point of this exercise is to notice your own imagination. There is no horse, no field, no fence, no sky, no meadow; these things do not exist, except in your mind. Yet, by the end of this exercise, most people will feel that they have "been somewhere" and seen these things.

Reflections

Do you find this kind of exercise pleasurable or does it seem like complete nonsense? Is it easy for you to form these images? Or is it something you have never done before, don't want to do again, and find difficult to achieve to your satisfaction?

Try these activities with a few friends and see what they think. Explore how your impressions compare and discuss their responses to these kinds of instructions. Ask them where in their lives they might use (either playfully or seriously) this kind of process.

Also ask yourself the same question. When might this kind of exercise be helpful to you in your regular activities? Think about a broad range of opportunities: from planning your day to planning your life; from organizing your room to building a city; from maximizing the use of a cafeteria to designing a new food distribution system.

The Seeds of Reality

By deliberately creating images you can sow the seeds for your own reality in all sorts of interesting and practical ways:

- Learn to anticipate – the view around the next bend in the road, the city that will be developed in the next decade, the faces in the audience after you tell your great joke.

- Learn to take your mind off the present – imagine a world that is better than the one you live in, or invent something that doesn't exist.

- Learn to extend your awareness – explore what you know about a friend's face without looking or try to trace out the different routes you could take to your house; try to experience the emptiness you would feel if a friend died, in order to better appreciate your relationship today.

- Learn to increase your flexibility – find a new way to solve a particular problem; develop a different perspective on an issue; vary and expand the routines you follow in your daily activities.

- Learn to rehearse your craft – to be ready for the swim race when the gun goes off; to have a first draft of an important memo ready before you hit the word processor; to anticipate all the rebuttals from your boss when you ask for a raise; to commit a speech to memory so that you can deliver it without using any notes.

To get a better sense of what it's like to use perceptual imagery, try the Back Side exercise in WarmUps. In this exercise, you're shown the front side of an object (such as a telephone, or clock) and asked to imagine the back side. 780

Will your anticipations, understandings, new perspectives, and rehearsals be exactly the same as the things that "really happen" to you? Maybe not. Will they prepare you for when things "really happen"? Almost definitely. Will they be useful? More than likely – especially if you can work at becoming comfortable with your spontaneous imagery and direct it effectively when real-time situations occur.

Perceptual Imagery and Conceptual Imagery:
Drawing and Diagramming

Depending upon your daily activities and profession, you may need to create highly detailed pictorial imagery as a part of your life. If you are an architect designing a house, it's important to imagine the building from a number of perspectives. If you are a product designer, you need to imagine different kinds of physical details and to consider objects in both whole and sectional views. If you are an interior designer, you need to imagine how people will use a kitchen space in order to figure out which way the cabinet doors should open. If you are a filmmaker, you need to imagine what landscapes, interiors, and interactions among characters will look like as a camera pans across them.

You even need a certain amount of imaging ability to set up a new computer system so that you can envision how you'll reach all the cabling after everything is installed.

Perceptual imagery can give the skills to represent a new idea and opportunities to yourself. To show these to others, and to help yourself in evaluating them, you will find that your drawing skills will be important. They will let you show explicitly what "objects of the world" you are suggesting.

And yet your imagination goes well beyond your ability to imagine new or recalled objects in the world. Your imagination is also critical in considering more abstract and conceptual elements. It is important in understanding new relationships between things and in inventing general approaches to new situations. In these instances, diagramming becomes your critical tool.

Exercising Your Imagination

Your imagination is multifaceted. It is perceptual and conceptual, specific and general, precise and fanciful. It is critical for providing you with a fresh view of your world.

You can exercise your imagination, making it more facile and effective. An exercise on the *CD-ROM* can enhance your imagination in a number of different ways.

Each of these approaches can cultivate your imagination in a slightly different way. Some will help you to generate perceptual images of new thoughts and ideas. Others are critical for encouraging you to use your imagination often and to be fluid with ideas. All will hopefully be fun enough to convince you to take more time to find more ways to use your imagination in your personal and professional life.

Imagining Spatial Relationships

Experimental research shows that some people are very good at perceptual imaging tasks, while others

find them impossible. Indeed, a great deal of folklore, as well as some research, suggests that individuals are either basically verbally oriented or visually oriented. Some people like to write down directions about how to get somewhere, for example, while others prefer to see a map. Some remember words easily and find themselves imagining in words – internally hearing and recalling great speeches, music, or arguments – while others remember the tiniest visual details and find themselves forming complex visual images in their spare time.

If you get particularly interested in spatial tasks or you find that you need more practice, you should look for puzzles in book or computer stores to extend your experience.

Because of this large difference, these kinds of tasks are often used to discriminate among people.

For instances, you might remember a number of perceptual tasks from the intelligence or aptitude tests you took in school. You may have either loved or hated these tasks, In either case, it's likely that you never had any formal training for these kinds of tasks. In fact, the only time you ever saw them was probably on tests.

Perhaps you extended your experience with spatial imagining by exploring puzzles, books, or games. You may have applied similar skills to carpentry projects, or, if you were a boy scout, you probably learned and applied information about spatial relationships by reading maps and calculating directions and learning about knots.

Transforming

In the TRANSFORMATIONS exercise on the *CD-ROM*, you can extend your ability to visualize spatial relationships.

Here your task is to look at a shape, and then imagine what it will look like after it is spatially transformed, for example, by a flip or a rotation.

Take some time with these exercises. Pay attention to your scores to see how you are doing. Use a timer to make the exercises even more challenging. Talk with friends to discover how they accomplish these exercises. Try to develop some explicit strategies for yourself, such as verbalizing a single feature, or paying attention to only critical features and ignoring the rest. Use your hands to mime the directions if this helps. Also try rotating the squares and cubes "in your head," assessing and developing your visualizing abilities. These are all ways to become efficient. Discover what works for you.

Reflections

The hardest part of designing TRANSFORMATIONS was developing a way to show the movements of the objects. For simplicity, we tried using words – such as "rotate right," or "flip horizontal," or "rotate left and flip vertical." Yet even we became confused by this! Did "rotate left" mean a 90° or a 45° rotation? Did "flip horizontal" mean flip it right to left *on* the horizon line or top to bottom *over* the horizon line? Different computer graphics programs interpret a horizontal flip in various ways, and people familiar with these were even more confused.

Take an object from your environment – a coin, a coffee cup, a wooden block and imagine what it would look like:
Flipped left
Flipped left and rotated right
Rotated right and flipped vertically

Ultimately, we decided that textual instructions were too ambiguous. The most difficult operation was reading the directions while imagining a rotation. For most of us, these activities didn't mix; we couldn't read and rotate together. It was so frustrating that it actually provided a good example of the differences between visual and verbal processing in spatial interpretation. We hope that the visual instructions we've added are clearer. Were they obvious to you? Do you think you might have preferred the verbal labels? Consider what other directions we might have used for this task.

As stated earlier, research seems to indicate that people do have a predisposition for either verbal or visual approaches. While some users enjoy solving the Transformations exercises and do so with a certain amount of playfulness, others are intimidated just looking at these puzzles, and feel overwhelmed at the thought of completing them.

Some people seem to be able to rotate images mentally as clearly as if they were holding the object in their hands, but most of us apply various strategies for reducing elements and identifying key features to make predictions.

Even if you're naturally more verbally than spatially oriented, practicing with these examples should help make these tasks easier for you. You'll learn to store images in your mind to predict the rotations, and you'll gather strategies that enhance your spatial predictions.

Building Forms

You can extend your three-dimensional experience by trying the BLOCK BUILDER exercise on the *CD-ROM*. This activity asks you to imagine how a large, multi-block structure might cast shadows. It requires that you take away cubic pieces of a large cube to create particular projections.

If you are like most people, you will find this task rather overwhelming at first glance. It requires not just single flashes of insight, but a deliberate study of multiple spatial relations.

Try it anyway. Once you take the first step of separating the block into different layers, it's not that difficult. Concentrate on one layer at a time.

Begin with the simple goals provided, then invent your own goals.

621 Block Builder Introduction
622 Try to Build Some 3-D Patterns
623 View Some Examples

You can change the game's preset shadows thereby setting your own goals simply by clicking on them. A favorite goal is to find a way to cast the shadows of your monogram (for example, if your name were Victor I. Zimmerman, you would try to show "VIZ" with the blocks). Note that you may need to be very clever about the shapes of certain letters in order to make them work for all three dimensions. Also note that some goals, that is, certain configurations, are impossible in this exercise.

Interpreting Descriptions

Transformations and BLOCK BUILDER exercise your perceptual imagination. They are very literal and precise, requiring that you see spatial relationships in your mind and predict changes.

But you can also generate imagery that is much more conceptual and impressionistic – images that are more like artist's sketches than photographs.

A wonderful way to generate this kind of imagery is by listening to words. If you let your imagination go with them, even single phrases can evoke incredibly rich images. If you're attended a poetry reading, you already know this.

Although some verbal descriptions are very literal, providing you with the particular details of a scene or object, most descriptions are more impressionistic. By not supplying the fine detail, they allow you greater freedom and control, encouraging you to change the images, sharpen the details, add color, assign specific features, and so on.

You can experience this simply by listening to the radio. A whole generation lived vicariously through the mystery, action, and adventure tales created for early radio. Clever narration, sound effects, and music enhanced listeners' experiences, helping them imagine the monster coming down the hall by the clomping of his footsteps, or the murderer approaching a victim by the hissing of her breath. Music built suspense, action, or romance as its pace quickened and slowed.

Many of us don't miss the imagery of television when listening to stories on the radio. We are pleased by the opportunity to control our own internal imagery, and to cultivate various interpretations of what's going on by using our "mind's eye." This experience is not the same as seeing, but it evokes many of the same feelings.

Imagine:
You are standing in a foyer. It is dark and cool, with only a few beams of sunlight touching the floor... 631

Ask yourself:
Where does the light come from?
What kind of floor is it?
631

Guided Fantasy

You can conjure up all sorts of objects, places, and events just by listening to audio descriptions. Using your own memory, you can imagine things you have seen before. Using your capacity for fantasy, you can also imagine objects and experiences that are not possible in the world as we know it.

As you listen to words, you'll find your imagination supplying you with a wide range of details. It's interesting to note that the amount of detail you generate may change radically, depending on how much attention you pay and what experiences you draw upon.

Your imagination is your passport to many interesting possibilities and conversations. It is not the same as seeing real things in the real world, but can help you see these things in new ways. It allows you to create new realities from "old" realities and to look at the world with a flexible mind.

An Environment to Imagine

WALKTHROUGH offers you the experience of being guided by a narrator's instructions and descriptions as you walk through an unseen house.

Notice the levels and kinds of details you form in your mind and how you rely on past experiences to create physical spaces and objects. You may have once known a similar house, and it may be supplying you with many of the visual images. Or you may have seen such a house in a movie, in a magazine, or even in your dreams.

Listen to this narration more than once to solidify your impressions. Try varying your images by changing certain aspects of the scene. Imagine, for example, a variety of different people living in this house: a stockbroker with her husband and two children; a Hollywood director and his starlet girlfriend; two college students; an elderly woman and her dog. Each of these inhabitants influences how the rooms are used and the objects they contain.

Now imagine this house being worth $50,000 or $200,000 or $3,000,000. Or vary the quality of light that falls on it by imagining different times of day, or different seasons in the year, or different types of weather. As in the imaginary horse exercise, notice how rich, diverse, and detailed are your images of something that, in reality, does not exist – and how much control you have over them.

Imagining Specific Details

Artifacts provides you with a greater amount of perceptual information than Walkthrough. It gives you the details and asks you to construct objects from them, rather than giving you a space and asking you to fill in the details. You may wonder whether these details were drawn from real objects, but it doesn't really matter whether they were or weren't (and we're not telling!). There are no right or wrong "guesses" to make in this activity. Your task is simply to imagine any of these objects as solidly and completely as you can.

---

631 Guided Fantasy Introduction
632 Imagine a Walk Through a House
633 Consider What Object Is Described
634 Enjoy Some Poetic Imagery
635 Create a Scene to Match the Sounds

" This object is...sort of drum-shaped and is made of dark, plum-colored heartwood. It has a lid and is a tiny sculpture of two sleek, plum-colored seals lying head to toe." 632

You might try drawing the objects you've imagined in your *Sketchbook*, alone or with friends. Compare your interpretation with theirs.

Try describing some of the objects in your own environment, making recordings of these descriptions for others to hear. As you do this, you'll probably be surprised at how clearly you have to see something in order to describe its details – and how difficult it is to decide which aspects to emphasize. You'll also learn how language can be used effectively to create images in other people's minds.

Poetic Imagery

Poetry and prose can be wonderful sources of inspiration for your visual imagination. Haiku, a form of Japanese poetry, is particularly renowned for the way it uses a handful of words to convey multi-layered ideas. Most haiku contain two contrasting or complementary ideas, a suggestion of the season, and a mood. Deceptively simple, they reveal (or evoke) more depth each time you listen to them.

Experience a few of the poems in the HAIKU section. Dwell on them, savor them. Slow down. Nurture an "in the moment" experience with your own imagination – rather than thinking about the next thing you can click on. Relish the visceral nature of the images you create. Enjoy their ambiguity. Delight in their simple provocativeness.

Worlds: Interpreting Fragments of Sound

An audio presentation need not be a verbal narration; it can also be a sequence of sound effects – like a radio drama soundtrack without a script. Try WORLDS to experience this. Listen to these varied noises and permit yourself to interpret freely what is happening. The images you create may seem less like drawings than a movie of a story developing before (or more accurately "behind") your eyes.

Relax and let each story unfold as certain possibilities emerge for you. If you compare your impressions with those of friends, you'll probably find a wide range of interpretations.

"What is really happening here?" you might ask. Guess what – no one really knows. These sound effects have been deliberately composed to convey a scenario, but they do not document an "event in the world" that has actually occurred. Welcome to the realm of imagination, where possibilities reign rather than single answers and interpretations!

Extend your experience of this activity by writing a story to accompany the soundtrack. It is script writing, or movie-making, in reverse; you start with the background, and must now place the action, the characters, and the motivation. See how many different stories you can create based on the same piece of audio.

Find other haiku poems that evoke your imagination, and try creating your own haiku. Analyze why certain images and approaches resonate for you. Do certain text phrases elicit almost a light-headedness, as you allow images to rush into your mind?

66 Artists treat facts as stimuli for imagination, whereas scientists use imagination to coordinate facts." -Arthur Koestler

The process of generating these imaginary worlds can be pleasurable and entertaining. Yet the opportunities go beyond this. We can construct these worlds to try out new things, to consider new possibilities; by doing so, we can learn about our own values and perspectives.

What If?: Planning, Designing, and Simulating
Our ability to consider objects that are not present and to make plans based upon them is a critical human activity.

Imagination is key to this ability. It allows us to tap into memories of objects and events we have experienced, and to extrapolate from these objects and events in the future.

One way we can augment our imaginative powers is by translating imagined future scenarios into concrete forms. These forms can then lead to collaborative frameworks in which we can consider and evaluate alternate possibilities. Such concrete, visual representations are particularly valuable in situations where the proposed changes are expensive, risky, or impactful. Acting as prototypes, these representations also allow us to initiate cycles of reflection and action.

The architecture/urban planning environment offers one notable arena for this approach. A number of urban design computer program provide visual simulations of cities, complete with a range of realistic variables.

The simulations allow you to imagine, for instance, what would happen to the crime rate if you introduced industrial centers into a residential area and moved the police departments to the city center. Or what would happen to traffic flow if you built additional housing around the suburban fringe or established a new arena for football games at the edge of the bay.

Based upon the alternatives you select, the program calculates and constructs the city that would result according to one underlying model. In this way, you can imagine a whole range of "what ifs" and "trade offs" and gauge their impact on the city you are designing. These concrete visual representations also create a framework that invites critique and collaboration from your colleagues.

Donald Appleyard and Ken Craik implemented an even more visual approach to scenario building in the Environmental Simulation Project at UC Berkeley in 1972. It was a project I had the privilege and pleasure of working on as a postdoctoral fellow. Appleyard and Craik had noticed that most architectural presentations offered a very abstract view of a building to potential clients. Though the presentations were highly visual, they typically showed the roofs of buildings (in models) or static views with model people (in sketches). However, a problem existed because human beings experience buildings and towns dynamically – by walking through them or driving around them – not by looking at models or sketches.

To address this, Appleyard and Craik developed a camera system that moved through and around architectural models. This "bird's eye view" representation gave non-professionals – the people who lived on the street, the client, the politicians – a more direct experience of the architect's plan. It helped them imagine the possibilities, and provided a vehicle for collaborative thoughts, discussions, and decisions.

I helped to create a similar "imagining tool" for the "Aspen Project" at the MIT Architecture Machine Group in 1978. The purpose of this project was to let people experience a place before they visited it – in this case, Aspen, Colorado. It differed from Appleyard and Craik's project in that it tried to represent a place that already existed rather than ones that were being planned, yet for the viewer this didn't matter. The Aspen Project was based on a series of interactive film clips, delivered on videodiscs; these "surrogate travel experiences" acted as "maps" of the area, providing highly visual data from which you could extrapolate and experience the town. Like the Appleyard and Craik project, it helped people imagine possibilities and served as a foundation for conversations.

Technologies have evolved a good deal during the past 20 years, since these demonstration projects were constructed. They are now extremely good at showing possibilities, constructing future scenarios, and relating experiences. They provide important capabilities for presenting and sustaining imagination in practical contexts.

Brainstorming: Generating Ideas

Where will we find the ideas to solve all of our problem? How will we move ideas and images out of our minds and onto the table (or up on the board) for conversation and analysis?

One well-developed method for this is brainstorming, a technique that is designed specifically to help people germinate new ideas and transfer these ideas into public dialogues.

A brainstorming session encourages free expression of ideas, yet it also involves certain constraints. It is one of those wonderful situations in life where freedom and creativity emerge best from clearly defined limits.

One of the important constraints for brainstorming is a time limit. Before a brainstorming session is organized, you should be precise about how long the activity will continue. Is this a ten-minute brainstorming session? An hour? A full day? It's critically important that there is a fixed time, otherwise people get nervous about the number of new ideas that are being generated, and they turn to the old ideas to be safe.

66 It is better to travel hopefully than to arrive."
-Japanese proverb

It's also important that all participants agree to the rules of the session. This ensures that everyone will feel included. It also encourages constructive participation and minimizes the chances that individuals will feel manipulated. Once the rules are established, new ideas will typically emerge from among the constraints.

Here are some of the rules that keep brainstorming sessions on the right track:

Go for quantity. Brainstorming sessions are an ideal forum for generating a large number of ideas. Just record the ideas as they come. Later discussions can explore controversies and provide analysis.

Defer judgement. Hold off criticism for awhile. As each idea is raised, it needs to be put up on a whiteboard or other display. Then, instead of spending time critiquing this idea, you should add a new thought. These sessions are not a place to be practical or to consider all implications. That can come later.

Get wild. It's better to generate ten wild ideas, even if you later discard them, than to be careful and come up with five ordinary ideas. The germ of a great idea is often found in a goofy, off-the-cuff suggestion.

Using a Facilitator

The rules of a brainstorming session are typically maintained by a facilitator, someone who is paying attention to the structure of the overall meeting rather than the specific topics at hand. Typically a facilitator introduces the rules from the very beginning, varying them according to the situation and the requirements of the session. Some brainstorming sessions work well without a facilitator, but most benefit greatly from a single-minded arbitrator who has an outside perspective.

The Brainstorming Process

Interactions in an effective brainstorming session are congenial, constructive, and even silly. People are not critical of one another, but collaborate toward the development of new ideas. The sessions are a series of quick-paced exchanges. By the end, most people have forgotten just who contributed what, and the resulting ideas are combinations of individual contributions. One idea evokes another. One person's concept encourages and even blends into another's. In the best sense, brainstorming is like child's play, eager and fun, yet with serious consequences.

66 If I were to sit down and come up with ideas on my own, maybe I'd come up with ten or fifteen ideas and it would take me half an hour. But with a group we can probably come up with fifty ideas in that same amount of time." -Larry Shubert

Interestingly, many people feel guilty spending time in a brainstorming session "when there is real work to be done," for the very reason that it can be so "fanciful" and so much fun.

But this attitude is terribly short-sighted. For one thing, you can often gain unusual insights on "serious work" by stepping outside a problem for a while. It gives you the opportunity to see things in a fresh way, and to notice patterns that are not accessible from "up close." For another, a problem often benefits from multiple perspectives, drawn from the talents of a variety of people.

We should note, however, that if you spend all your time brainstorming, you are not going to get the rest of the work done! This process is wonderful for harnessing your imagination and generating new ideas. But it needs to be used judiciously, with some discipline. Take the "gems" these sessions reveal and then plan activities around them. Without follow-through and analysis, the process can be fun and generative but not necessarily productive.

It's true, too, that brainstorming sessions are more appropriate at certain times during a project. All stages of a project do not call for the same degree of flexibility. Sometimes you just need to get your head down and work out the details of what's already proposed, letting the ideas evolve in a convergent direction.

The general consensus is that brainstorming is most effective during the early phases of a project, when you are beginning to formulate the task, when you are particularly open to many possibilities, and when you are engaged in "problem finding" activities.

But if you give the principles of the ARC cycle serious consideration, you can see that brainstorming is useful not only at the onset of projects, but at the beginning of every creative cycle that the project goes through! As you move from action to reflection, you can garner critique, and as you move from reflection to action, you can gather new ideas.

Finally, be aware that if your work-flow process gets stuck, or when it seems to have reached a premature end, a brainstorming session can often break the pattern. "Necessity is the mother of invention," and often the most bleak situation can be resolved by a "group think."

Try It Yourself

The only way to learn how you can personally benefit from brainstorming is to try it. There are a number of diverse exercises on the *CD-ROM* for you to try out, by yourself, with a friend, or in a small group.

Try brainstorming on these topics:
Design a house for a rabbit
What would you do with 12 pounds of peanuts?
List ten new ways to keep rain off your windshield

66 Brainstorming is always cool because I might be working on a project that's in the medical field, and learn about a new technology that I can apply to a bicycle product. Crossover technologies are really exciting; when you get people from all different backgrounds coming together... it's really very synergistic." -Lynette Ross

As you begin to address these tasks, be very clear about the rules you will follow. Assign yourself 10 (but no more than 20) minutes on each topic. Follow the basic rules of brainstorming (or your version of these) and make sure everyone understands them.

After you brainstorm ideas for a topic, take a few minutes to analyze your productivity. Think about what worked well and what didn't. Try to improve your skills. Identify the different abilities and perspectives each of you brings to the sessions and analyze which are effective.

Be surprised. Try out some silly ideas. See where you get.

Once you become comfortable with brainstorming, you may wonder how you could have ever been afraid of running out of ideas. You will no longer think of yourself as a "person with no imagination."

As you experiment with it, you'll discover that brainstorming is applicable to a wide range of topics and people. It is equally important for the product designer who is inventing a new toothpaste tube; the mom who is planning a birthday party; the scientist who is challenging a theory; the second grader who is writing a story; or the museum designer who is putting together a new exhibit.

Creating Sequences

Imagination is critical for storytelling. Great story-tellers know how to structure a story to give it form and how to embellish it to add to the mystery and delight as it evolves. Filmmakers know how to engage an audience in a story, using images, music, and sound effects to capture the viewers' own emotions and imaginations. They excel at getting each of us to "suspend our disbelief" for a few hours as we allow the images on the screen to become a personal reality.

Storytelling is an inherent human trait. We tell stories almost every day of our lives, when we recount what happened at work, how last night's party went, or how we solved a problem. The difference between merely recounting a true experience or crafting an engaging story from the same experience lies in the degree of imagination you bring to it.

The storyboard format is a common one among designers, filmmakers, animators, and comic book artists. They use it to think through a series of actions or angles, or to portray the progression of an idea, or to tell a story in separate but linked visual stages. Complex movies typically use extensive storyboards to help the director, model builders, photographers, and set designers decide what they want the audience to see and how. These sketches are often recreated down to the last detail on film.

651 Introduction to Magic Theater
652 Make a Story with Three Given Images
653 Create a Story Around Nine Images
654 Your Turn to Select All Your Images

Create your own storyboards off-screen, using pictures cut out of magazines or your own photographs. Scatter these at random and pick them up one at a time, as in Stage Two. Or pick a few out of the pile and then link them to others, as in Stage One. Write text to go with each image, or just talk through the storyline as it unfolds.

In the MAGIC THEATER exercise on the *CD-ROM*, you can exercise your ability to tell tales in a storyboard format by specifying sequences of images and sound effects. You can encourage the development of the viewer's understanding by weaving together ambiguous imagery and sounds.

Try this exercise. Explore STAGE ONE and STAGE TWO first, since they have some constraints that may challenge and inspire your imagination.

Learn to "just give it a try" and don't worry about getting the "right answer." Go with the flow of your ideas, no matter how preposterous or improbable they may seem. This is a playground for your own inventiveness. Enjoy!

Ambiguity

A deliberate feature of the MAGIC THEATER is the ambiguity of its images and sounds. They were rendered to encourage multiple and metaphoric interpretations. The dice, for example, do not necessarily represent literal objects; they can symbolize fate or luck, or a range of different things. "So destiny rolled the dice one day," you might say, or "Life's a crap shoot," or "I thought about all the times I had gambled on love and lost."

Wide shots of hillsides can signify important personal journeys, the expressions on people's faces can appear happy or sad depending on the context you give them. The sound effects are equally flexible, allowing you to pair them with images in ways that enhance your story.

This ambiguity should help you, as a storyteller, to tell the story you want. Your ultimate goal is that your audience see and understand your interpretation. A sequence of images is not in itself a story; the story is created for the viewer, who is skillfully influenced by words and sounds (and the storymaker's vision) to look at the images in a particular way.

We hope you have fun with this exercise. You may be surprised to discover that you are full of ideas for a range of stories. Some stories can give you enormous satisfaction, and reveal a great deal about your mind and memories. Others are just silly or absurd pastiches of images and sounds.

Process, Not Product

You may be disappointed to find that you cannot save your work. A completed story is erased when you begin another or when you go to another section on the *CD-ROM*. We debated this decision considerably as we designed this section, for in the process each of us had made some stories that we wanted to keep and share with our friends.

Creating a story in front of a friend or a group can make the whole process very amusing as your storyboard evolves.

We concluded, however, that the activity was not so much about the stories we created, but the process of imagination that went into making them. We could always reconstruct a particular story; in fact, subsequent attempts would probably lead to improvement rather than simple replication. Imagination is a fluid thing. The MAGIC THEATER is designed to encourage your confidence in your imagination; you will be able to invent many products after you become comfortable with this process.

If you have ever observed children building a sand castle on the beach, you will have noticed that they are equally delighted by the eventual destruction of the castle by encroaching waves as they were by the process of building it! I have watched my own children laboriously construct tall towers with blocks, then giggle with delight as they knocked them over.

Unlike children, most adults emphasize the products in their lives. Products represent our accomplishments and are our evidence of hard work. Yet, at their core is the exercise of our imaginations. As adults, we need more opportunity to play just for the pure joy of it, unencumbered by the burdens of final "deliverables."

So there is no "save" function in the MAGIC THEATER. It has been deliberately constructed with a fancifulness and impermanence that should encourage you to play. And if you really fall in love with a particular story, you can always write down the numbers of the images and sounds as well as the text phrases!

Play Is Serious Business

Although we have focused largely on its visual aspects, your imagination is multidimensional. It allows you to predict visual changes in the world as well as to consider alternate realities. It enables you to conjure up both detailed and impressionistic images in your mind and to bring forth ideas into public arenas for discussion and analysis. It helps you link diverse or unrelated items and to present them to others in meaningful forms.

Many imagination activities are fanciful, even silly, and just plain fun. As a result, they are the activities that we've been encouraged to leave out of our lives when we are involved in "serious business." They are the mental playgrounds where we seldom go, unless we're retired, on vacation, independently wealthy, or paid specifically to spend time imagining and creating.

But this can, and should, change.

Imagination is one of the priceless treasures in life that everyone can afford.

Enjoy your imagination. Use it to see, draw, and diagram fluently. Use it to tie these activities together, to identify issues, to solve problems, to dream of unrealized (but achievable!) possibilities. Engage your imagination in an action-reflection-change cycle as you see what is and imagine what might be. Become part of the visual culture in which personal imagination can be both public and collaborative, without diminishing its power.

Putting It All Together

Imagination tops off your visual abilities. Your environment and the methods of the visual culture are at the base of your abilities; they provide the foundation for your activities. Seeing, drawing, and diagramming provide you with the ability to take advantage of what's around you, bringing your perceptions and thoughts into presentation and conversation.

The *VizAbilities* cube has provided a framework for you to explore each of these perspectives on the intersection of your mental and physical worlds. You have begun to understand these opportunities, and to develop the skills that let you take advantage of them.

Now it is time for you to combine all of these perspectives and benefit from their interactions. Begin with your imagination and then make a drawing. Diagram in order to reflect on a problem, working within the ARC cycle of the visual culture. Gather inspiration by really seeing your environment. Rearrange your environment by drawing. Grab a new idea to encourage your imagination from what you see in a diagram.

Combine the singular perspectives of the faces of the cube to benefit from their interaction. Consider the solid form of the cube as the representation of your integrated visual abilities.

Epilogue:
Visuals in the Information Age

Writing at the turn of the century, Francis Galton suggested that images were part of a primitive sort of thinking that was found frequently among women and children, but not in learned men. The same kind of "you'll grow out of it," attitude is inherent in many recent developmental theories. These theories assert that humans move from concrete reasoning to abstract reasoning, for example, implying that image-based thinking is a state to be moved beyond.

As we evolve dynamic electronic tools to extend our visual abilities, it is time to abandon such a limited view. We need to extend our abilities to think and reason with all the elements available to us – including movies, sketches, storyboards, drawings, diagrams, and photographs – rather than to constrain ourselves to a primitive concept of thinking and reasoning that is based on text and the spoken word.

The development of written language was a magnificent achievement. It allowed us to preserve our culture and to extend it. It gave us the ability to communicate great thoughts and deep emotions, to tell wonderful stories. The printing press enhanced our ability to disseminate these records. Its invention and the development of an educational system that enabled us to learn how to read and write these arbitrary written symbols brought the power of literacy to most of the world.

Yet, by themselves, these achievements are not enough. We need to include imagery among our everyday skills – not just as passive observers, but as active visualizers. We need to incorporate visualization techniques in our informal dialogues, our public collaborations, and our formal presentations. We need to couple this powerful tool with our verbal skills so that we can address the increasingly complex nature of the modern world.

Imagine for a moment that, instead of the printing press, Gutenberg invented media-rich computers that could produce, store, and display movies, sounds, and text – and that these elements could be sent anywhere in the world within seconds.

How might our cultures have evolved?

For one thing, we probably would have maintained our oral traditions. We probably would have nurtured the art of direct conversation, and preserved fluid conversations as records rather than pages of text. We would have emphasized dynamic representations (and maybe moved more quickly toward understanding mathematical ideas, such as calculus and chaos theory, which are inherently dynamic). We would have trained ourselves to use imagery in our daily interactions, employing it to send ideas over distances and make records of everyday occurrences.

But Gutenberg imagined the printing press instead, and imagery took a back seat to the alphabet. It is only in this century that images have begun to reveal their worth again and that technologies have allowed this form of expression to reach a multitude of diverse people.

We have created media-rich computing. We are evolving image-based global communication networks. We now need to anticipate the cognitive capabilities these technologies will both require and enhance. We need to create a culture that can tap into and exploit all these capabilities.

Our visual abilities are key to the creation of this culture.

Of course, your "vizabilities" can develop quite independently of computers. In most instances, drawing and diagramming are still more easily accomplished on paper than on computers. Seeing typically requires only your eyes. Imagining can happen anywhere, with no tools at all.

Yet, as new computer systems evolve, we'll all have even more opportunities to think and describe in visual terms. Activities that are currently complex and therefore in the hands of professionals or the talented elite will become accessible and mainstream. As computing devices get smaller, and networking capabilities get less awkward, we'll be able to send and receive images as rapidly as we now transmit words through the air.

And the digital revolution – where all modes of representation, visual or not, are integrated in the same format – promises to streamline this visual language into an accessible, everyday mode of communication.

We can build on this opportunity, creating a visual culture that takes advantage of evolving technological tools. We can shift the destiny of images from concrete forms that live on the pages and screens of other peoples' works, to tools that we can all use to communicate our ideas.

We have barely begun to explore how computing devices can "augment the human intellect," as Doug Engelbart, the inventor of the mouse and a pioneer in designing computer systems for human use, suggested decades ago.

In the classic book *Flatland,* the flatlanders lived in two-dimensional worlds guessing at the three-dimensional worlds surrounding them. In our everyday lives, we live without much sense of the dynamic and interactive potential that surrounds us, though we are learning more and more about it from its projections into our current activities.

We must continue to develop our visual vocabulary, to work and think comfortably with dynamic and interactive representations; and to be ready to join the digital world and to become familiar with its unfolding potential.

We are in the process of inventing a new human language, one in which *you* can now participate by recognizing and utilizing your own natural *VizAbility*.

References

ABBOTT, EDWIN. *Flatland.*
New York: Barnes and Noble, 1963.

ADAMS, JAMES, *Conceptual Blockbusting.*
San Francisco: W.H. Freeman, 1974.

ARNHEIM, RUDOLPH. *Art and Visual Perception.*
Berkeley: University of California Press, 1954.

ARNHEIM, RUDOLPH. *Visual Thinking.*
Berkeley: University of California Press, 1969.

BARLOW, HORACE; BLAKEMORE, COLIN AND WESTON-
SMITH, MIRANDA. EDS. *Images and Understanding.*
Cambridge University Press, 1990.

CHING, FRANCIS D.K. *Drawing, a Creative Process.*
New York: Van Nostrand Reinhold, 1990.

CLAY, GRADY. *Close-Up: How to Read the American City.*
New York: Praeger Publishers, 1973.

D'AMELIO, JOSEPH. *Perspective Drawing Handbook.*
New York: Van Nostrand Reinhold, 1964.

DE BONO, EDWARD. *Lateral Thinking.*
New York: Harper and Row, 1970.

DE BONO, EDWARD. *Teaching Thinking.*
London: Penguin Books, 1978.

DEWEY, JOHN. *How We Think.*
Boston: Houghton Mifflin, 1910.

DONDIS, DONIS. *A Primer of Visual Literacy.*
Cambridge, Mass.: MIT Press, 1973.

DEWEY, JOHN. *Democracy and education.*
New York: The Free Press, A Division of Macmillan
Publishing Co., 1944.

E. GOMBRICH. *Art and Illusion: A Study in the Psychology
of Pictorial Representative.* Princeton, NJ: Princeton
University Press,1960.

EDWARDS. *Drawing on the Right Side of the Brain.*
New York: St. Martins Press, 1979.

ENGELBART, D.C.. *"A Conceptual Framework for the
Augmentation of Man's Intellect."*

Eyewitness Books: Sports.
New York: Alfred A. Knopf, 1988.

GORDON, WILLIAM. *Synectics.*
New York: Harper and Row, 1961.

GREGORY, R.L. *Eye and Brain: The Psychology of Seeing.*
New York; McGraw Hill, 1966.

HALPRIN, LAWRENCE. *The RSVP Cycles: Creative Processes in
the Human Environment.* New York: George Braziller, 1969.

HANKS, KURT AND BELLISTON, LARRY. *Rapid Viz: A New
Method for the Rapid Visualization of Ideas.*
Los Altos, CA: William Kaufmann, Inc., 1977.

HOWERTON AND WEEKS (EDITORS). *In: Vistas for
Information Handling.*
Washington, D.C.: Spartan Books, 1963.

Images, Objects and Illusion Readings from Scientific American.
San Francisco: W.H.Freeman, 1974.

KIM, SCOTT. *Inversions.*
New York: W.H. Freeman, 1989.

KIM, SCOTT. *The New Media Magazine Puzzle Workout.*
New York: Random House, 1994.

KLEE, PAUL. *The Thinking Eye.*
New York: George Wittenborn, 1961.

KOBERG, DON AND BAGNELL, JIM. *The Universal Traveler.*
Los Altos, California: William F. Kaufman, 1972.

LANGER, SUZANNE. *Feeling and Form.*
New York: Charles Scribner's Sons, 1971.

LASEAU, PAUL. *Graphic Problem Solving for Architects and Builders.* Boston: CBI Publishing, 1975.

M., CZIKZIMIHALYI. *Flow: The Psychology of Optimal Experience.* New York: Harper and Row, 1990.

MCCAULEY, DAVID. *Castles.*
Boston: Houghton Mifflin, 1977.

MCCAULEY, DAVID. *Cathedral.*
Boston: Houghton Mifflin, 1973.

MCCLOUD, SCOTT. *Understanding Comics.*
North Hampton, MA: Kitchen Sink Press, 1991.

MCKIM, ROBERT. *Experiences in Visual Thinking.*
Boston: PWS Publishing, 1979.

MULLET, KEVIN AND SANO, DARRELL. *Designing Visual Interfaces.* Englewood Cliffs, NJ: Prentice Hall, 1995.

OPPENHEIMER, FRANK. *Working Prototypes: Exhibit Design at the Exploratorium.* San Francisco: Exploratorium, 1986.

ROBIN, HARRY. *The Scientific Image: From Cave to Computer.* New York: Harry and Abrams, Inc., 1992.

SHEPARD, ROGER. *Mind Sights.*
New York: W.H. Freeman, 1990.

SIBBET, DAVID. *Group Graphics: Empowering Through Visual Language.* San Francisco: Grove Consultants, 1981.

SIBBET, DAVID. *The Facilitation Guide: Principles and Practices.* San Francisco: Grove Consultants, 1993.

SIBBET, DAVID. *Fundamentals of Graphic Language.*
San Francisco: Grove Consultants, 1992.

TUFTE, EDWARD. *Envisioning Information.*
Cheshire, Connecticut: Graphics Press, 1990.

TUFTE, EDWARD. *The Visual Display of Quantitative Information.*
Cheshire, Connecticut: Graphics Press, 1983.

SEYMOUR COHEN, LUANNE; BROWN, RUSSELL; JEANS, LISA AND WENDLING, TANYA. *Design Essentials.*
Mountain View, CA: Adobe Press, 1992.

WOLFF, ROBERT AND YAEGER, LARRY. *Visualization of Natural Phenomena.* New York: Springer-Verlag, 1993.

WHORF, BENJAMIN *Language, Thought and Reality.*
Cambridge, Mass.: MIT Press, 1956.

WURMAN, RICHARD SAUL. *San Francisco Access*
New York: Access Press, 1987.

 011

 101

 122

 111

 123

 112

 112

 124

 113

 131

 114

 132

 121

 133

These images are the jump screens from the *CD-ROM*. Pressing the Command (for Macintosh) or Control (for Windows) key followed by the three digit number indicated next to each screen, will allow the user to jump directly to these locations on the *CD-ROM*.

 Environment 101

Public Space
Objects
134

Studio
Introduction
141

Studio
Views
142

Studio
Visit
143

Studio
Objects
144

Workspaces
Gallery
152

Culture
201

Sketching
Introduction
211

Sketching
Ideas League
212

Sketching
Whiteboards
213

Sketching
Gallery
214

Prototyping
Introduction
221

Prototyping
Views
222

Prototyping
Visit
223

Prototyping
Objects
224

Critiquing
Introduction
231

Critiquing
Views
232

Workspaces
Introduction
151

Culture 201

 233

Critiquing
Process

 234

Critiquing
Game

 241

Readiness
Introduction

 242

Readiness
Relax

 243

Readiness
Focus

 244

Readiness
Listen

 245

Readiness
Reflect

 301

Seeing

 311

Flash Sketching
Introduction

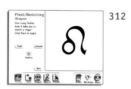 312

Flash Sketching
Shapes

 313

Flash Sketching
Images

 314

Flash Sketching
People

 321

Eye Tracking
Introduction

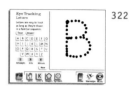 322

Eye Tracking
Letters

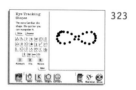 323

Eye Tracking
Shapes

 324

Eye Tracking
Symbols

 331

Hidden Pictures
Introduction

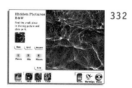 332

Hidden Pictures
B&W

 Seeing 301

 333

 341

 342

343

344

 351

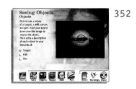 352

 353

 354

 355

 401

 411

 412

 413

 421

 422

 423

 424

 Drawing 411

 431

 432

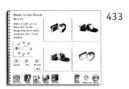 433

 434

 441

 442

 443

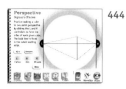 444

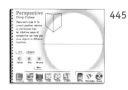 445

 451

 452

 453

 461

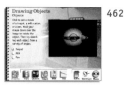 462

 463

 464

 465

 501

Diagramming 501

Using Diagrams
Introduction
511

Using Diagrams
Dialog
512

Using Diagrams
Group
513

Using Diagrams
Present
514

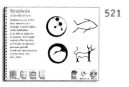

Symbols
Introduction
521

Symbols
Match
522

Symbols
Draw
523

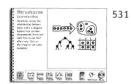

Structures
Introduction
531

Structures
Linear
532

Structures
Cluster
533

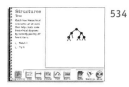

Structures
Tree
534

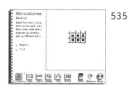

Structures
Matrix
535

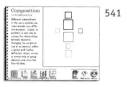

Composition
Introduction
541

Composition
Position
542

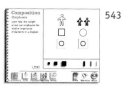

Composition
Emphasis
543

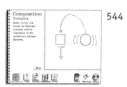

Composition
Grouping
544

Show and Tell
Introduction
551

Show and Tell
Map
552

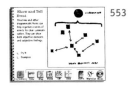 553

 554

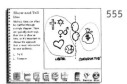 555

 601

 611

 612

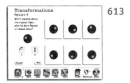 613

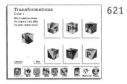 621

 622

 623

 624

 625

 631

 632

 633

 634

 635

 641

 Imagining 601

 642

 643

 644

 645

 651

 652

 653

654

 710

 720

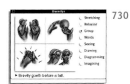 730

 740

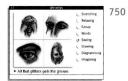 750

 760

 770

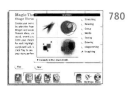 780

Colophon

This book was designed and produced at MetaDesign's San Francisco studio using Adobe Illustrator, Adobe Photoshop, Equilibrium Debabilizer, Microsoft Word, and QuarkXPress on Macintosh computers. Proofs were printed on a Hewlett-Packard LaserJet 4MV and a Canon CLC 300 with an Adobe PostScript RIP. Final film was output at 150 LPI on a Scitex Dolev 450 image-setter, by Rapid, San Francisco, California. Printed by Quebecor Press, Kingsport, Tennessee.

Typefaces used were Clarendon designed by H. Eidenbenz, 1953; ITC Officina Sans designed by Erik Spiekermann, 1990; and Swift designed by Gerard Unger, 1985.

The following individuals or groups contributed original photography, videography, illustration or animation to this product: Ted Casey, CMCD, Gayle Curtis, Richard Haukom and Associates, IDEO Product Development, Image Club, Scott Kim, MetaDesign, Kevin Ng, Corinne Okada, Dale Pedersen, Cindy Rink, Deskin Rink, Kristee Rosendahl, Ross Elementary School, David Sibbett, Gerry Slick, Wendy Slick, Stanford University (ME-101 Visual Thinking), Oenone Terrill, Paul Trachtman, Annie Valva, Jym Warhol, Kristina Hooper Woolsey, and Sabrina Yeh.

Drawing by Dürer courtesy of Phaidon Press Ltd.